"Friendship is a very comforting thing"
A. A. Milne

Happy Birthday Bruce!!

Love,
Diana

6/8/13

The Wine Book
Wines and Wine Making Around the World

ALEXANDER DOROZYNSKI
BIBIANE BELL

The Wine Book

*Wines and Wine Making
Around the World*

Prepared under the direction of
LOUIS ORIZET

Recipes by MADELEINE OTHENIN-GIRARD

GOLDEN PRESS NEW YORK

Contents

Preface	11
THE ORIGINS OF WINE	14
FROM VINE TO WINE	42
FRANCE	68
Bordeaux	69
Burgundy	93
Champagne	111
Loire Valley	122
Rhône, Provence, Midi	131
Alsace and Lorraine	139
Jurançon	144
Arbois and Jura	146
Corsica	147
WESTERN EUROPE	150
Italy	151
Germany	167
Switzerland	184
Austria	189
Spain	191
Portugal	199
Belgium and Luxembourg	206
NORTH AFRICA, GREECE AND THE NEAR EAST	208
North Africa	209
Algeria	209
Tunisia	211
Morocco	212
Greece	215
Israel	218
Lebanon, Syria, Jordan	221
Egypt	222
Turkey	223
Iran	224
EASTERN EUROPE	226
Soviet Union	227
Hungary	234
Bulgaria	240
Rumania	241
Czechoslovakia	241
Yugoslavia	242
UNITED STATES	249
LATIN AMERICA	265
Argentina	266
Chile	267
Brazil	268
Uruguay	270
Peru	270
Mexico	271
SOUTH AFRICA	271
AUSTRALIA	274
WINE APPRECIATION	278
SOME RECIPES	293
Map: Wine Production Throughout the World	308–309
Acknowledgments	307, 310
Index	311

Illustration facing title page:
"THE YOUTHFUL BACCHUS,"
painting by
Michelangelo Amerighi Caravaggio,
1595. *Uffizi Gallery, Florence.*

Published in 1969 by Golden Press, New York, N.Y.,
a division of Western Publishing Company, Inc.
© Copyright 1968 by Les Editions des Deux Coqs d'Or,
Paris, and Mondadori O.G.A.M., Verona.
Designed by Michel Duplaix.
Library of Congress Catalog Card Number: 69-19127
Printed in Italy by Arnoldo Mondadori Editore - Verona

PREFACE

There are thousands of books on wine, and it is not the intention of the authors simply to add yet another encyclopedic compilation of the best-known growths. Wine growing is a way of life and this is a book which attempts to examine the various countries that produce wine in the light of their wine-growing and wine-drinking traditions. The picture is different each time. The wines of one country or area have certain characteristics that set them off from, or resemble, those of others. But characterizing and appraising these wines is a tricky matter and should only be undertaken with care.

One could say, for instance, that it is doubtful that the viticultural technology and wine-growing trends of certain Slavic countries will ever lead to the production of wines to the taste of Western Europeans or Americans. Does this mean the wines are poor? Such a statement would be presumptuous, for there are laws of relativity in matters of wine, too. Nothing can illustrate this better than a personal experience. As a member of a jury during an international wine competition, I was particularly severe in my judgment of Bulgarian wines. In the isolation provided for jury members, I had nothing to go on except the wine, my own senses, and the memories of wines I had appraised in the past. These memories, fortunately, happen to be those of prestigious Médocs and unforgettable wines of Burgundy's Côte de Nuits. In the monastic setting of the jury room, the Bulgarian wines simply did not measure up to the same standard, hence my severity. Two days later, however, I was invited to a Bulgarian gathering and dinner, where the same wines were served. My attitude changed quickly. The wine's excessive vigor, even harshness, that had shocked me in the jury room took on a completely different aspect when served with Bulgarian dishes. In the gaiety of the evening, I could not but admit my bias. This lesson, following so many others, has rendered me cautious with respect to myself, indulgent with respect to others.

The liking or disliking of a particular wine or type of wine is an intensely personal matter, and is influenced by physiological, psychic, even religious factors. In Scandinavia, for example, the taste runs to heady wines, rich in alcohol and sugar. One can only conclude that the basis is physiological, having to do with the cold climate. Similarly, during the war years in France, when food was scarce, the consumption of extremely sweet wines increased, doubtless as a means of providing calories lacking in the daily ration.

WINE BOTTLE. *Musée des Arts décoratifs, Paris.*

But taste in wines is more than physiological. It is formed by our backgrounds and histories, our passions and prejudices, the very tenor of our lives. And as our civilization becomes more and more one of leisure, esthetic considerations overcome physiological needs. The primary, functional aspect of wine—unlike that of other agricultural products such as rice, potatoes, or wheat—tends to recede. Wine in much of the world has become an added comfort, a flourish of life. France, for example, is oriented toward the selection of vine plants encouraging the production of subtle wines, in which such esthetic considerations as taste, aroma, and a clear and pleasing color dwarf such practical matters as strength, quantity, or sweetness.

In the United States, where nutritional shortcomings are not a problem, the trend seems to be in the same direction. America has inherited a bewildering assortment of wine-growing traditions from the world over and is in the process of nationalizing them, adapting them to its own needs. It is likely that an original wine-growing tradition, and probably an interesting one, will develop. Borrowing at first from European traditions, such wines as California Burgundies, Chiantis, and Clarets are beginning to develop their own qualities and, progressively, their own proper names. As they become more individualistic, they will proudly wrap themselves in their own banners and quietly drop—like the toys of childhood—the names and traditions that gave them their first impetus.

I hope that readers will be interested in knowing about the wines, the customs, and the techniques of wine growing throughout the world, as they emerge in the chapters devoted to them. The approach to wines in this book is unavoidably Western and its judgment cannot help but be influenced by our Western traditions and way of life. These are reflected not only in the wines that we make and like to drink, but also in opinions expressed about the wines discussed.

Louis Orizet
Inspector General,
French Institute of Controlled
Place Names

WINGED SPIRIT with a human body and eagle head
sprinkling the sacred tree with lustral water. Assyrian bas-relief from Nimrod.
9th century B.C. *Louvre, Paris. (Photo Bulloz)*

The Origins Of Wine

FOSSIL VINE LEAF. A sample of *Vitis balbiani,* the oldest known variety. Sorbonne geological laboratory, Paris. *(Photo Jacques Boyer-Roger Viollet)*

THE VINE SINCE PREHISTORY

It is unfortunate for the chroniclers of human events that the first sip of wine has been swallowed up in the murky twilight of prehistory. The story of wine is inseparable from that of man. A knowledge of the drinking habits of our early ancestors (whose alcoholic beverage was undoubtedly some sort of wine) could have added interesting clues to the process of human evolution: wine, a powerful emotional catalyst, is known many times to have swayed the course of history.

Such a study, alas, must be confined to more or less reliable conjectures. Vinous fermentation is a natural and spontaneous phenomenon that paleontologists date back at least a hundred million years to the Mesozoic era, the time of the brontosaurus, the tyrannosaurus, and other huge reptiles.

Long before man, earth was covered with luxuriant vegetation—notably the lianalike ancestor of modern *Vitaceae*, the large family of plants of which the wine-bearing *Vitis vinifera* is a member. Several of these precursors have survived to this day, such as the European *Ampelopsis*, a sturdy plant that was tried, unsuccessfully, as a graft-bearer for decadent or diseased modern varieties.

The genus *Vitis*, which includes all of the domesticated vines, appeared 30 or 40 million years later, during the Eocene times of the Cenozoic era—the age of mammals. Fossil vines 500 million years old have been found in the region of Champagne: *Vitis sezannensis,* a plant very similar to several modern varieties of the southeastern United States, was a robust rampant, but almost certainly could not bear wine (claims made by Champagne enthusiasts notwithstanding).

The Pliocene period, a dozen or so million years before man, saw the emergence of many new kinds of *Vitis:* for example, in France, the *Vitis ausoniae* (named after the Latin poet and wine lover Ausonius), a juicy grape almost identical to modern *Vitis vinifera,* and the wild *Vitis labrusca,* whose progeny was used for wine making by early American settlers. *Vitis labrusca* still grows in much of the United States east of the Mississippi.

The stage was set for man to step upon the grape.

Each anthropologist may have his own axe to grind when it comes to deciding who was the first "true man," but few doubt that fermented beverages from fruit or berries were known to the earliest manlike creatures more than 100,000 years ago, whether he was the *Australopithecus* or *Zinjanthropus* of Africa, or the *Sinanthropus* of China. It is a foregone conclusion that man's first acquaintance with wine was an accident: grapes gathered and hoarded by our ancestors needed only the prompting of natural yeasts to turn into a wine of sorts. The reader may fancy the facial expression of the low-browed, big-jawed *Zinjanthropus* when he discovered that the grapes he had stored away in a rocky ledge for a rainy day had fermented: disappointed, perhaps, at the sight; perplexed by the taste; stunned, no doubt, by the effect; but, almost certainly, hugely pleased.

Man's acquaintance with wine gradually became more purposeful. It is likely that Cro-Magnon man, some 30,000 years before our era, civilized and leisurely enough to be an accomplished decorator of his cave dwelling, found time to study and exploit the mysterious phenomenon of fermentation. European lake-city dwellers of Neolithic time some 8,000 years ago certainly made wine—a conclusion suggested by packed grape seeds discovered in their garbage dumps. These fossilized dumps indicate that Stone Age man had a varied menu, including bread, soups, shellfish and fish, as well as assorted meats, and it is not far-fetched to wonder whether he was already selecting his wine to fit his meal.

A few anthropologists have even ventured the opinion that the need to make wine, rather than to grow food, gave rise to the first Stone Age agricultural settlements, for Neolithic people still fed chiefly upon the proceeds of hunting.

THE MAGIC BEVERAGE

That wine became associated with the very first religious manifestations is a generally accepted hypothesis. Primitive religions took shape when man turned to supernatural forces to account for awesome events whose cause was not readily known: fire, lightning, storms, disease, and death—and to the exhilaration or stupor that befell one who partook of alcoholic drink. Intoxication was a form of magic that lent itself particularly well to ritual, for it could be provoked at will, unlike such supernatural manifestations as thunder and lightning, which could hardly even be predicted.

It is not surprising, then, that wine, as well as fire (another somewhat controllable form of magic), has been involved in many religious rites, from ancient Babylon to modern times. Nor is the opposite astonishing—that wine and alcohol in general came to be forbidden by some religions as a manifestation of evil spirits.

Yale anthropologist Donald Horton notes that "the individual inebriate is conspicuous by his absence from most primitive communities." Wine was relatively scarce, and its drinking was rather a collective occupation limited to religious festivals or to those occasions when it became necessary to communicate with the powers that be. Wine drinking could even be delegated to a selected member of the group who entered the "holy state" on behalf of his brethren.

In patriarchal societies (which have been more frequent than matriarchies in the history of mankind), women were often forbidden to drink, as they were considered altogether unworthy of it. Wine—red wine of course, reminiscent of blood—is also a traditional offering to the gods. The earliest known cuneiform writing, on tablets discovered at Erech on the Persian Gulf more than five thousands years ago, mentions that a sort of wine was used for sacramental purposes.

The Bible has more than five hundred references to wines, vines, or vineyards, beginning with Noah:

Noah was the first tiller of the soil. He planted a vineyard; and he drank of the wine, and became drunk . . . —Genesis 9:20-21.

NOAH'S DRUNKENNESS. A miniature from the Paris, or so-called *Rohan, Book of Hours,* about 1418. *Bibliothèque Nationale, Paris. (Photo Bibliothèque Nationale)*

usq; donec veniens staret supra
ubi erat puer. Videntes autem
stellam: gavisi sunt gaudio ma
gno valde. Et intrantes domum
invenerunt puerum cum ma
ria matre eius. Et procidentes
adoraverunt eum. Et apertis
thesauris suis: obtulerunt ei mu
nera. aurum thus et mirram.
Et responso accepto in sompnis
ne redirent ad herodem: per ali
am viam reversi sunt in regio
nem suam. Pater noster. Et
ne nos. Sed libera nos amalo.

According to a Mesopotamian text discovered at Nineveh in 1962, wine was stored among the provisions for Noah's Ark: "For our food, I slaughtered oxen, and killed sheep day by day, and with beer, oil and wine, I filled large jars."

EARLY VINEYARDS

It was in the region of Mount Ararat, now shared by Iran, Turkey, and Armenia, that Noah would have sighted land; and it is here that man is believed to have started the first large-scale cultivation of the vine. As early as the fourth millenium B.C., large quantities of wine were stocked around dwellings. Great jars were unearthed at archaeological sites of that period, bearing traces of fermentation and containing dried-out wine deposits. Large paved areas used for trampling the grapes were found nearby.

From the slopes of Mount Ararat, viticulture spread to the fertile valleys below. Recent Soviet archaeological expeditions in Turkmenistan, at Ak-Tepe and Naurazga-Tepe, have unearthed North Iranian cuneiform texts referring to the wine trade. At Ivriz, north of Tarsus, there is a huge Hittite bas-relief of the second millenium B.C. of a man, believed to be a god, holding grapes in his hand. Sumer, a civilization of the lower Euphrates which antedated that of Egypt, was also well acquainted with wine.

In Egypt, wine had been known as far back as the fourth or fifth millenium B.C. and is depicted on scores of surviving records of religious and artistic activities. Many jars and frescoes show the primitive methods of making wine from grapes: long lines of chariots drawn by asses carried the harvest in woven baskets, the grapes were trampled by foot or wrung out in cloth, and the juice channeled for fermentation into earthen jars.

Osiris, son of the sky and of the earth, was also "master of the flowering." During the religious celebrations, wine was given in offering to the gods, and some was also distributed to the population, which ordinarily had to settle for beer, palm-tree wine, and *shebdu,* a heady drink brewed from pomegranates. Wine was included in funerary offerings to accompany the dead, and it was used in preparing the body for its long voyage—the entrails were removed to be cleansed with wine and then replaced.

If the use of wine was restricted to an elite, this elite, it seems, often abused its privilege. Many frescoes show scenes of drunkenness. A description in a

AMPHORA BEARERS.
Fresco from the tomb of
Serekhotpe, Thebes,
about 1450 B.C. Tracing.
British Museum, London.
(Photo Chapman)

hieroglyphic text of the Seventeenth Dynasty (circa 1600 B.C.) tells of a woman requesting eighteen bowls of wine. "I love drunkenness," she adds. A fresco shows a woman casually turning away from a table to vomit into a bowl held by a servant—a practice which enabled a reveler to go on drinking far beyond his or her capacity. Egyptian soldiers received rations of wine, and priests had an abundant supply: records show that priests during the 31-year reign of Ramses III (1198–1167 B.C.) received 722,763 jars of wine and jugs of beer.

THE EMPEROR AND THE JADE GIRL

In China, several legends have it that wine was invented under Emperor Yu, some 4,000 years ago. Emperor Yu reigned during the time of an eight-year-long war and disastrous floods, and worried so much about the fate of his subjects that he could neither sleep nor eat. But one day Yi Tieh, a maid in his palace, served him wine she had prepared and, after tasting a cup, the Emperor found his long-lost appetite had returned. He drank more with his meal, and his worries vanished.

Another legend goes on to explain that the art of making wine was in fact imported from a distant "Western Paradise" by the Jade Girl, one of the Emperor's concubines. What went into the making of the emperor's wine is not known, though the reference to the Western Paradise hints that it might be the grape. Another story has it that wine was only introduced from the Caspian Sea region during the first Han dynasty, in the first or second century B.C.

At any rate, whatever Chinese wines were made of, abundant records confirm their existence. During the 2,000 years preceding the reign of the fourth emperor of the Yuan dynasty, about A.D. 1300, prohibition laws against the use of wine were established and repealed no less than 41 times. Confucius (circa 500 B.C.) preached moderation: "Drinking knows no limit, but never be boisterous with drinking."

His Indian contemporary, Buddha, was more strict: "Never take strong drink," he ordered in one of his commandments. In the Indian tradition there is mention of wine in the Vedas, sacred books in which the Soma represented the sacred beverage. The identity of Soma, however, is a mystery; the fermented juice of at least a dozen plants has been suggested as the ingredient.

Closer to the birthplace of wine, Persia (Iran) has had its famed wines of Shiraz in the Farsistan—praised by Hafiz and Omar Khayyam—for more than 4,000 years. Sumero-Akkadian documents firmly establish that the early Mesopotamian people drank much wine, and pre-Islamic Arabs in the deserts wrote poetry about it. One bit of old Arabic wisdom has it that "the wineskin is a kingdom to him who possesses it, and the kingdom therein, though small, how great it is!"

In regions of hot, arid summer, the vine plants were extraordinarily vigorous. The Greek geographer Strabo reports that there existed, near the oasis of Margiana (today Mary in Soviet Turkmenistan), vines with trunks as huge as trees, bearing bunches two cubits (about three feet) long.

GRAPE HARVEST AND WINE MAKING. Fresco from the tomb of Rameses, Thebes, about 1456 B.C. Tracing by Nina de Garis Danes. *British Museum. (Photo Chapman)*

WINE IN THE BIBLE

In Palestine, the vine played an increasingly important role from the second millenium B.C. on. Around 1400 B.C. Moses dispatched scouts to the Promised Land. Two of them arrived in Canaan.

And they came to the Valley of Eshcol, and cut down from there a branch with a single cluster of grapes, and they carried it on a pole between two of them . . .—Numbers 13:23.

They brought this fruit to Moses, and told him it came from a land of milk and honey.

The Old Testament frequently points out the importance of viticulture in the history of the Jewish people. When Jews migrated to Chaldea on the Euphrates river and the Persian Gulf, they took along vine plants and contributed to the spreading of viticulture throughout Mesopotamia. Palestine also knew prohibition, according to Jeremiah, who arrived at the house of the Rechabites and generously set before them pots full of wine and cups, and invited them to drink the wine.

OIL LAMP, showing a scene from the Bible: Joshua brings back the grapes from Canaan. Terra-cotta. *Lavigerie Museum*

JUG IN FORM OF GRAPE CLUSTER. Black glass with polychrome neck. Egyptian, 18th dynasty. *Louvre. (Photo Chapman)*

But they answered, "We will drink no wine, for Jonadab the son of Rechab, our father, commanded us, 'You shall not drink wine, neither you nor your sons for ever; you shall not build a house; you shall not sow seed; you shall not plant or have a vineyard; but you shall dwell in tents all your days, that you may live many days in the land where you sojourn.'" —Jeremiah 35:6–7.

Those ascetic tent dwellers, however, were the exception rather than the rule. The Book of Proverbs strikes a more sympathetic note:

... it is not for kings to drink wine, or for rulers to desire strong drink: lest they drink and forget what has been decreed, and pervert the rights of all the afflicted.

Give strong drink to him who is perishing, and wine to those in bitter distress; let them drink, and forget their poverty, and remember their misery no more.—Proverbs 31:4–7.

Philologically, one can now trace wine through the longest span of time with the greatest of ease. In 1500 or so B.C. the Hittites, whose language was then predominant in Asia Minor, referred to wine in their cuneiform script as *wee-an*, in hieroglyphic as *we-anas*. This seems to be related to the Sanscrit *vena*. The oldest known Greek name for wine is found on an archaic inscription as *woinos*, an obvious ancestor of the classical Greek *oinos*, and relative of the Latin *vinum*, the Italian (and Russian) *vino*, the Georgian *gvino*, the Hebrew *vayin*, the Sabaean *wayn*, the modern Arabic and Ethiopian *wa-yn*.

Everywhere wine is closely associated with religion. First it occurs in violent rites during which wine was mixed with blood—as in the frantic "mysteries" in honor of the Persian God of Light, Mithra. Then, as humanity slowly yielded to civilization, many of these periodical blood-baths were replaced by wine baths, and libations or wine offerings took the place of sacrifices—a custom evoked in Homer's Iliad at the meeting of King Agamemnon and Odysseus. "They drew wine from the bowl in cups, and as they poured it on the ground they made their petition to the gods that have existed since time began." Wine became the god's blood which men could drink to reach outside of themselves toward a divine being—an inspiration common to nearly all the world's religions.

DIONYSUS

The Greek wine god Dionysus, son of Zeus, was a prolific contributor to folklore, and wine played an important part in the flowering of Greek civilization. Born as a horned child crowned with serpents, Dionysus took many guises during his life, appearing as a bull, a ram, a stag, a kid, a lion or a tiger, as a girl, a handsome youth, or an old man. The Greeks have represented him under a variety of forms—most frequently as a strong, tall, bearded figure, often crowned with ivy, or as a beardless, delicate youth. Only seldom (until the Roman Bacchus) do we see the God of Wine as a bibulous, obese, jellylike decadent, weakened by alcoholism.

Legends of Dionysus' travels vary, but it is said that he first sailed to Egypt, accompanied by a wild army of Satyrs—sharp-eared, ram-footed musicians, and Maenads—redoubtable women wearing foxes' pelts and wielding ivy-twined staffs. He was well received in Egypt, where he planted wine before returning to Greece. There he established the wine cult.

It was, at first, a violent one. Under the blazing torchlights, Dionysian followers, coated with goatskins, like living wine flasks, and Bacchae, wearing the Maenads' fox pelts, reveled in secret orgies held to the sound of the flute (Dionysus' favorite musical instrument) and the accompaniment of off-color incantations. These drunken Dionysian orgies frequently ended in bloodshed. (Orpheus, the gentle poet, was torn limb from limb because he had neglected to honor Dionysus.)

Then, slowly, the refining influence of Hellenic civilization toned down the savagery of Dionysian rites. The tradition of orgiastic celebrations persisted longest in Thebes and Delphi, while elsewhere the festivities became more bucolic, taking place in September during the harvest, or in January, when the mysterious fermentation had to be propitiated or else demons might take over and turn the wine sour.

BACCHANAL.
Greek bas-relief.
Prado, Madrid.
(Photo Anderson-Giraudon)

Some of the Dionysian reunions became the occasion for theatrical competitions between rival groups, giving strong impetus to classical Greek theater. During the 5th century B.C., the state organized national and regional wine festivities, the great Dionysia and the urban Dionysia, in Icaria, Athens, and other great cities, where the masterpieces of Greek theater were born. The wine god became the patron of music and dance—indeed, the patron of all arts. Dionysus joined the gods on the frieze of the Parthenon, displacing Hestia, the Goddess of Earth, so that the number of major gods remained twelve.

Theatrical representations were consecrated to Dionysus; his bust presided and his priests sat in the front row. And only after the play was there another competition, more directly connected to the activities of the patron god—a race to empty a large cup filled with new wine.

SYMPOSIA

Parallel to the official Dionysia, the symposium, a more intimate "drinking together," became fashionable among the intelligentsia and the *bons vivants*. Such a symposium, unlike its counterpart of today, was not a gathering of scholars or scientists to discuss learned matters, but a drinking party, spiced with more or less elevated conversation.

Speaking at a symposium related by Plato, Socrates is quoted as saying: "Gentlemen, insofar as drinking is concerned, you may have my approval. Wine moistens the soul and lulls our griefs to sleep while it also awakens kindly feelings. Yet I suspect that men's bodies react like those of growing plants. When a god gives plants too much water to drink, they can't stand up straight and the wind flattens them; but when they drink exactly what they require they grow straight and tall and bear abundant fruit. And so it is with us. We pour ourselves great draughts and before long our bodies and our minds reel and we talk nonsense; but if the servants constantly keep filling small cups only, we are brought by the gentle persuasion of the wine to a gayer mood."

Even ascetic Plato did not oppose moderate drinking of wine and merely declared that "to drink to the degree of drunkenness is not becoming anywhere, except perhaps in the days of the festival of the god who gave men wine for their banquets." By 400 B.C., social drinking, moderate or immoderate, had come to stay. "Let us drink, for this is the very truth that wine is a horse for the road, while foot travelers take a stony path to Hades," aptly observed Antipater of Tyre, a stoic philosopher.

Athenians held symposia to feast a god, celebrate a victory, greet a returning friend or bid a departing one farewell or, simply, to get together with a few congenial guests. Symposia were not, as a rule, the gatherings of many. Half a dozen friends, almost always men, were enough to keep a conversation going without letting it become dispersed.

Dinner was served first, and once the food was out of the way, the tables were removed, a libation poured to the gods, and a butler or "inspector of the wines" prepared the vintages to be drunk during the evening, usually adding some water to them. A floor show was frequently provided, a favorite subject being the enacting of an encounter between Dionysus and Ariadne, whom Dionysus had married after she had been abandoned by Theseus.

There was a genuine concern for the proper ways and means of drinking during these occasions. Plato quotes Pausanias as admitting at the outset of a symposium that he was still breathless from the drinking of the previous night, and as asking: "Well, gentlemen, what will be for us the most comfortable way to drink?" The participants, including Socrates, then start a lively discussion, which is interrupted by the arrival of Alcibiades, a cheerful, dissipated youth, whose political activities contributed to the downfall of Periclean Athens.

Alcibiades empties a small bucket of wine, then hands another to Socrates. "This is not a trap," he comments, "as no matter how much Socrates is told to drink, and no matter how much he drinks, he is never drunk." At this point, everyone starts sipping the wines. This particular reunion ended up by Socrates putting his companions to sleep, washing up and strolling over for the rest of the day to the Lyceum's shaded, peaceful walking paths favored by Athenian poets and philosophers.

PERSIAN MINIATURE, 16th century. Note the vine spiraling about the tree.
British Museum. (Photo British Museum)

SATYRS DANCING. Canthare, Kabirion (Thebes). *Louvre. (Photo Chapman)*

THE KOTTABOS GAME

Social wine drinking in Greece led to the invention of a curious game, *Kottabos,* which remained highly popular for at least three centuries. The game, combining grace with skill, consisted of striking a collapsible target with the small amount of wine remaining in a cup after the draught is drunk. The target, resting atop a rod, was usually a metallic saucer. If the aim was good, the target tipped over, to clatter noisily onto the floor or to strike another metallic object below.

It seems the game was invented by Greeks in Sicily, as a fitting entertainment after a meal. For one thing, it required no great physical effort that would disturb the digestion. For another, it allowed the company to admire the graceful motions of young people of both sexes, their heads covered with wreaths of flowers, taking careful aim and twirling their cup with a flick of the wrist to send a few drops of wine toward the target—called a *plastinx*. The game grew to be so popular that those who could afford it had their houses equipped with special *kottabos* rooms, circular in shape so that the couches of all the participants, lined up against the wall, were at the same distance from the central pole and the *plastinx*. Professionals gave instructions in the game, in which the grace of stance and motion were counted together with the accuracy of the aim. Special *kottabos* cups were made, their shape designed to favor the accurate delivery of the liquid.

YOUNG MAN
PLAYING KOTTABOS,
from a Greek painted amphora.
Louvre. (Photo Chapman)

On the sides of a wine cooler at the Hermitage Museum in Leningrad, there is such a scene of *kottabos,* played by a svelte girl, identified as Smirka, wearing no clothes and reclining on a couch while she takes careful, graceful aim. Another vase in the Louvre, Paris, shows a young man on a couch swinging his cup and, at the same time, raising his leg to counterbalance the motion in a stance somewhat reminiscent of that of a baseball player, if a baseball player were trying to pitch from bed.

A GAME OF KOTTABOS, from a Greek drawing.
"L'Imagerie" collection. (Photo S. Hano)

A good shot at the *plastinx* knocked it off the supporting rod with a ringing noise that signified a bet was won, or a prize was to be offered by the host. Among prizes, according to Sophocles, were kisses and love play. Sometimes the players wagered small objects, or their clothes—though often they didn't bother to wear any at all, since clothing was liable to hamper the accuracy of a throw.

HOMERIC WINES

The Greeks attached great importance to their wines. Jars were periodically inspected and tasted by cellar masters. The wine was carefully "bottled" into long-necked amphorae after spending several months in the jars. Heavy wines with a strong alcoholic content were kept for several years before being drunk— at least five years, a lapse of time respected for the best Bordeaux wines of our time. Some wines are reported to have been kept for as long as 150 years, though by that time they were reduced to a concentrated essence to be drunk with water or mixed with younger wines. Homer lyricized the aroma of wine as being "sweet and marvelous such as no one could resist."

When Odysseus sailed from Troy to Ismarus, famed for its wine, and sacked it, he destroyed the town and slew its men. Then, "from the town we took the women and great stores of treasure, and divided all, that none might go lacking his proper share. This done, I warned our men swiftly to fly; but they, in utter folly, did not heed. Much wine was drunk . . ." So much that the natives found

A DRINKING CUP.
Rhodes, Mycenaean civilization.
Louvre. (Photo Chapman)

MIXING BOWL.
Mycenaean civilization,
14th century B.C.
Louvre. (Photo Chapman)

time to regroup and to slaughter many of Odysseus men. The remainder of his crew "sailed with aching hearts, to be clear of death though missing our dear comrades." They did not sail, however, before loading the ship with "large stores in jars" of the famed Ismarean wine.

Odysseus' personal cellars were renowned. Large earthen jars were lined up against the walls; the oldest and the best wines were cautiously stored in the cellar's deepest vaults, while the wine for current consumption was more accessible. Wine was drawn from jars into large amphorae, which were still too big to be used at the table; it was served from smaller amphorae or from pitchers by slaves or wine boys, under the watchful eye of the cellar master, whose duty it was to make sure that all cups were kept full. During Odysseus' travels, his best wines were locked up in his room, together with his jewels, his best clothes, and his gold.

HELLENIC GROWTHS

There was a huge variety of wines. The poet Virgil once observed that "it would be easier to count every particle of sand in the sea than to list all the growths of Greek wine." The most famed growths were wines of Chios, one of the eastern islands of the Aegean and one of the presumed birthplaces of Homer, where the poet is believed to have lived for many years. From Chios came Arvisian and Phaenean wines, whose reputation was as high as their prices. One

KYATHOS, 5th century B.C.
Louvre. (Photo Chapman)

RYTHON. Rhodes.
Louvre. (Photo Chapman)

ETRUSCAN DRINKING CUP. *Louvre. (Photo Chapman)*

metrete (about ten gallons) of a good vintage from Chios fetched as much as one *mina,* the cost of two oxen, a price comparable to that of an exceptional French vintage today.

Other Aegean islands produced excellent wines. Lesbos, Samos and Lemnos had reputable growths, as well as Naxos, where, according to a legend, a natural wine used to flow from a spring. On the mainland, the city of Thrace, where Dionysus was said to have been born, held its own, notably with the Maronean wine which Odysseus, imprisoned in the Cyclops' cave, served to the one-eyed monster. The Cyclops became so drunk that Odysseus was able to pierce his eye with a charred tree-trunk, and to escape with his men. (The Cyclops' own "black wine," made from heavy clusters of grapes that grew profusely without being cared for, was considered a poor, barbarian beverage.)

Macedonia's Acanthian and Mendean growths were excellent, and Thessaly prided itself on its Heraclean growth from Thermopylae. But neither Attica nor its capital, Athens, apparently had much to boast about viniculturally. The memory of only one growth has lived through history; Chrysatikos, the golden wine, and even it, according to Alexander of Tralles, was artificial.

After the death of Alexander the Great, who drank a great deal of wine in his later years, his empire was divided among his generals. Greek power declined, and by 146 B.C. Greece was a Roman province.

BACCHUS THE GUEST OF ICARUS. Roman bas-relief, Naples. *National Museum, Naples. (Photo Alinari)*

WHEN IN ROME

The early Romans were frugal and drank little wine, but in later times they acquired the reputation of extremely heavy drinkers. Roman wines were less refined than those of the Greeks, and the Roman Bacchanalia haled back to the violent tradition long replaced in Greece by the gentler symposia. In the sacred woods near Rome, wine festivals were celebrated by torchlight, and often ended up in drunken brawls and slaughter. Legislators appointed special Senate investigating committees to look into Roman drinking habits. Many excessive revelers were arrested and some executed, and in 189 B.C. the *Senatus Consultum de Bacchanalibus,* a sort of constitutional amendment, was adopted to outlaw Bacchanalia in Rome and Italy.

It seems that the Romans nevertheless kept drinking wine rather heavily. A custom persisted for several generations of toasting the emperor during banquets by drinking a cup of wine for each letter of his name; this may have been healthy enough under Nero, but less so under Vitellius, and downright dangerous under Septimus Severus. (Of course even under Nero it could be disastrous if one wanted to use his full official name, which was Nero Claudius Caesar Drusus Germanicus.)

The Romans made their own wine, but much of it was mediocre, and patrician connoisseurs who could afford Greek vintages preferred to import them. It was perhaps because of the scarcity of good wines that Roman women were not allowed to drink wines—whereas Greek women (except in Sparta) shared this pleasure with their menfolk. A stiff penalty threatened the Roman housewife caught red-handed in her husband's amphora: "The husband is the judge and jury to his wife," wrote Cato the Elder (234–149 B.C.). "His power cannot be appealed; if the wife has misbehaved, he punishes her; if she drank wine or committed adultery, he kills her."

Polybius, a Greek historian, noted: "It is almost impossible for them (Roman women) to drink wine without being found out. For the woman does not have charge of the wine; morever, she is bound to kiss all of her male relatives and those of her husband down to her second cousins every day on seeing them for the first time, and as she cannot tell which of them she will meet, she has to be on her guard."

ROMAN VINEYARD. Mosaic found at Tabarka. Note the curious way of arranging the vines in spirals and at regular intervals. *Musée du Bardo, Tunis. (Photo Musée du Bardo)*

By the 1st century B.C., during the blossoming of Roman civilization and the golden age of Roman poets, Italian wines had improved. Virgil, Martial, Horace, Tibullus and Catullus speak of the wines of Palermo, Mount Massico, Cecubis, Alba and the wines from the "Seven Hills" around Vesuvius. Falerna was promoted to be the official wine of the Imperial Court, though Pliny the Elder, a student of natural sciences and connoisseur of good wine, wrote that Falerna was too strong and harsh, and advised against drinking it unless it was mellowed by ten to twenty years of aging. ("And I speak of wine with the seriousness of a Roman passionately interested in the arts and sciences, as a judge intent upon the salvation of mankind," added Pliny.)

Wine was in such demand that wine making became a serious, competitive business that spread throughout the Italian peninsula. Agronomist Columella in the 1st century A.D. gave in his *Treatise on Agriculture* practical advice to would-be winegrowers.

Seven *jugera* of vine was as much as a single grower could properly tend to, decreed Columella (a *jugerum* being equivalent to 28,000 square feet). Such a parcel of land, well situated for wine growing, could be purchased for about 7,000 sestertii, or approximately $300 to $400. The expense for plants, stakes and planting would run another 2,000 *sestertii* per *jugerum*. Then 8,000 *sestertii* should be earmarked for the purchase of a good slave to tend the wine, bringing the total investment for a vineyard of seven *jugera* (4 and ⅜ acres) to 29,000 *sestertii* or so. To this should be added interest of 6 percent for the first two years that vines did not bear fruit, for a total of 32,480 *sestertii* or roughly $1,500 to $2,000.

If the yield did not exceed one *culeus* per *jugerum* (a reasonable estimate, as the culeus equaled 20 amphorae, or 135 English gallons), a fair price would be about 300 *sestertii* per *culeus*, "the cheapest price" in Columella's book, "at which wine could be sold, and still bring a fair return of upward of 6 percent of the money invested." This would amount today to a few pennies a gallon. Of course the wine would have to be good, and not *vappa*. Vappa designated wine turned sour, and the word was jokingly used by the Romans to describe a worthless blockhead. (It was eventually deformed into the Neapolitan *woppo* and persists in American slang in the form of *wop*, a usually disparaging term for an Italian or one of Italian descent.)

FIRST GALLIC WINES

When Julius Caesar crossed the Alps into Gaul in the 1st century B.C. he found no vineyards—or at least he did not describe any in his account of the campaign, though he carefully noted other types of vegetation and terrain, and studied the way of life of the natives.

Thus it seems that before our era, the ancestors of the vintners of France— today the greatest wine-producing country—were not great wine makers or wine drinkers (even though archaeological finds indicate that the inhabitants of Gaul knew how to make wine). Their national drink, rather, was a sort of barley beer —a stupefying brew if one is to believe Aristotle, who says that "a man who is drunk with wine falls flat on his face, because wine makes his head heavy; a man who is drunk with beer falls flat on his back because beer stupefies." Aristotle was quite emphatic about the backward fall, though less about the forward pitch, for he writes that "under the influence of all other intoxicants, those who get drunk fall in all directions, sometimes to the left or to the right, sometimes on their faces or on their backs. But those who get drunk on barley beer (*pivos, zythos*, or *krithinon*) only fall backward and lie on their backs."

Phoenicians from Asia Minor probably sailed to the Gallic coast before the Greeks, but they did not trade in wine. It must have been the Greeks, then, who first brought the forward pitch to Gaul.

The exact time of the first Greek colonization of the Gallic "far west" is still debated. Hellenist Pierre Demargue believes that the first Greek scouting expeditions to southern Gaul and the east coast of Spain took place in the second half of the 7th century B.C. Massilla (today's Marseilles) was the first Greek base in Gaul, founded in about 600 B.C. by sailors from Phocaea, a Greek city of Ionia. The Greeks settled along the coast, apparently on friendly terms with the Ligurian natives.

After Cyrus' conquest of Asia Minor in 538 B.C., many more Ionian Greeks emigrated to the new colonies, establishing several ports, such as Monoïkos (Monaco), Nikai (Nice) and Antipolis (Antibes). They imported wine plants and used native vines to start wine growing on a scale previously unknown in Gaul. But wine growing was restricted to the coast where the Greeks lived—which may explain why Caesar did not observe any vineyards when he marched north.

The Greeks traded with the natives more than they warred with them, slowly expanding their influence, establishing new cities, or taking over ancient Phoenician trading centers. On the eve of the decline of Greece, the settlers had conquered much of the Western world—not by arms, but by commerce. The expansion of the Greek wine trade in Gaul is indicated by the finding of many typical long-necked Greek wine amphorae, even in central France.

THE FLOURISHING WINE TRADE

The Romans, once more, followed in the footsteps of the Greeks. Even before the Roman occupation of Gaul, the wine trade thrived between the Romans and their neighbors across the Alps. Wine was transported from Italy in flat, round sailboats called *corbitae*. One of these sank off the Gallic coast in the year 240 B.C. and was perfectly preserved on the muddy bottom until it was discovered a few years ago. The ship belonged to one Marcus Sestius, merchant. She was carrying Greek wines loaded in the island of Delos, and had stopped in Italy to load up with Latium wine. She held about a thousand amphorae and 800 pieces of assorted pottery; some of the amphorae had been closed and pitched so well that when they were discovered they still contained a yellowish liquid—the remnants of wine more than 2,000 years old.

The Greek or Roman wines were unloaded in seaports and shipped to customers inland, on muleback or in mule-cart caravans, or in barges pulled up-

GALLIC SHIP TRANSPORTING WINE. Bas-relief from
Trier Museum, Germany. *(Photo Chapman)*

river. Amphorae were not returnable, and thousands have been found, broken up, in garbage disposal pits and on the bottom of rivers.

Gradually, the Greek trade declined, and the Roman trade increased. After Caesar's conquest of Gaul, the commerce in wine, almost exclusively Roman, was boosted by an influx of new customers—Roman settlers.

The Gauls were now learning to make good wine themselves, and the Romans, fearing their profitable export business might collapse, forbade the natives to plant new vines on their soil. But the agile ancestors of the French promptly found a way to bypass this ruling. In order to plant their own vineyards, they bribed Roman soldiers or civil servants to buy a parcel of land, which then became Roman soil. It was perfectly legal to plant vineyards on Roman soil, and once the vineyards started to bear fruit, the parcel of land was resold to the natives. The law, which did not allow them to plant wine on their land, did not require the destruction of "existing vineyards."

The Gauls planted Italian and Greek vines, and also cultivated local varieties. Eventually the competition the Romans had feared became fierce, particularly in the region of Narbonne, which started shipping wine to areas heretofore exclusively supplied by Italian merchants. Around A.D. 90 Emperor Domitian ordered that about half of the vines in the Roman provinces be torn out. It is not exactly known why he took this drastic measure. Was it to favor Roman merchants? Or was it because wine growing had been taken up with such enthusiasm and on such a scale that it was seriously interfering with other crops?

The Domitian edict remained in force for nearly 200 years, until Emperor Probus gave the "barbarians" the right to make wine *ad libitum,* hoping this world insure their loyalty to Rome.

CHRISTIANITY

In the first centuries of our era, another force, more powerful than Greek fleets or Roman legions, moved the world: Christianity—which contributed to spreading the culture of wine even to those regions where the climate makes wine growing totally impracticable. Wine was a part of the Christian religion, playing a role that was, at first, no more than a new approach to the ancient symbolism associating wine and divine blood. Later it was to become considerably more important.

Christ's first miracle was the turning of water into wine at the marriage feast in Cana, Galilee:

Now six stone jars were standing there, for the Jewish rites of purification, each holding twenty or thirty gallons.

Jesus said to them, "Fill the jars with water." And they filled them up to the brim.

He said to them, "Now draw some out and take it to the steward of the feast." And they took it . . . the water now became wine . . .

This, the first of his signs, Jesus did at Cana in Galilee, and manifested his glory; and his disciples believed on him. —John 2: 6–9; 11.

Sanguis uvae, the blood of the grape, became a necessity for the followers of the new religion. Certain sects of the Primitive Church still exploited the mystic state reached through drunkenness, establishing a communion with God by entering ecstatic trances achieved by drinking consecrated wine. St. Paul himself was indignant and found it necessary to issue a warning against this dangerous deviation, reminding them of Christ's words:

"This cup is the new covenant in my blood. Do this as often as you drink it, in remembrance of me."—I Corinthians 11:25.

St. Paul also warned that:

Whoever, therefore, eats the bread or drinks the cup of the Lord in an unworthy manner will be guilty of profaning the body and blood of the Lord . . .

For any one who eats and drinks without discerning the body eats and drinks judgment upon himself.—I Corinthians 11: 27; 29.

HARVESTING GRAPES AND TREADING WINE.
Miniature from the Paris, or so-called
Rohan, Book of Hours. Bibliothèque Nationale, Paris. (Photo Bibliothèque Nationale)

SEPTEMBER. Fresco from the Pritz chapel,
Laval, 12th century. *(Photo Archives photographiques)*

The change of wine into divine blood remained an article of faith, but henceforth the faithful were not to seek communion with God through drunkenness —though there persisted an aftertaste. Witness these words of St. Theresa: "I consider the center of our soul as a cellar into which God leads us, when He is pleased by this admirable union, so as to inebriate us in a holy way, with this wine made so delicious by his Grace . . . All our senses are left at the door, as if asleep."

The spread of Christianity from the sunny shores of the Mediterranean caused the grape to grow far out of its natural environment. Isolated parishes inland had to grow their own grapes and make their own wine because transportation was too slow or too hazardous to provide a regular supply of wine for the mass. Wine making was of interest, also, because wine was wealth—a liquid asset that could be stored and whose value increased with time, like a savings account.

Priests became trained in the art of wine making. The custom was to surround a new parish, church or abbey with its own vineyards, and the suitability of the terrain for wine growing often determined the location of a new religious outpost. *Fecit ecclesias et plantavit vineas*—"He has built churches and planted vines"—significantly reads a diploma granted by Emperor Charlemagne at the end of the 8th century to a monk dispatched to establish a new monastery.

"Northern wines" came into being around churches as far north as Gloucestershire in England. Most of the northern vineyards have disappeared, but notable survivors today are those on the shores of the Rhine and the Moselle, many of which were planted on the orders of Charlemagne himself.

From the 15th century on, Christian Europe went off to discover, then to conquer, the world. Colonization, evangelization, and the planting of wine went together. Portuguese missionaries introduced wine to Japan and Brazil. Spaniards spread wine to the rest of South America, Central America, and California. French Protestants carried it to the Cape of Good Hope. And to the east, the expansion of Russian domination put at the disposal of the Orthodox Church wine-growing areas which had been Islamic, hence dry.

Of all the Christian denominations, the Catholic church, with its millions of followers in lands most favorable to wine growing, was probably the strongest force to propagate wine in the world. "Of all religions, the only true one, the Catholic religion, is the only one to exalt the wine and sanctify it, to eucharize it on its altars," preached Father Edouard Krau, curate at Vosne-Romanée in Burgundy a few years ago. "There is only Catholicism to valorize your wine . . . For Catholicism, wine production is a *sine qua non* condition of existence, for without wine, no mass, and without mass, no Church."

THE ANTIOCH CHALICE. Silver, Byzantine, 1st century.
Private collection. (Photo Giraudon)

From Vine to Wine

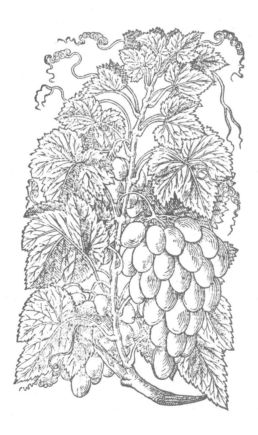

VINA PASSA. Wood engraving from the Mattioli manuscript, 16th century.
"L'Imagerie" collection. *(Photo Chapman)*

Preceding pages: PRUNING THE VINES.
Fresco by Francesco del Cossa. Ferrara, Italy. *(Photo Scala)*

THE WINEGROWERS

Many if not most people view the culture of the vine and the making of wine as an amiable pastime, better suited to please man and occupy his idle hours than to test his vigilance and perspicacity. Of the beautiful but difficult profession of winegrower, they know only the image of gay *vendanges* (grape harvests), with joyful bands of harvesters singing under the sun. Of the subtle art of making wine, they have a picture of the smiling cellar master, proffering the glass of friendship in the hospitable atmosphere of an ancient cellar.

Of course, all this is part of the life of a winegrower. But there is much more, too, where skill must restrain love—and courage yield to resignation.

The vine, after all, is a notorious rebel. She has never forgiven man for having pulled her from her ancestral forest to compel her to live, domesticated, regimented, lined up in serried ranks as if for a parade. Never completely resigned to this new way of life, the vine has many cunning ways of seeking revenge. Sometimes excessively exuberant, sometimes languishing, she has become a fragile, unpredictable member of modern society, capriciously rebelling against servitude by threatening not to fulfill the mission assigned her by her master.

Wine, her offspring, behaves just as erratically, having inherited a whimsical and wayward character. It seems that wine likes to be disconcerting, to baffle those devoted to its making. If man is willing to forgive this insubordination, it is perhaps because he finds in wine a mirror of his own weakness and changing moods, his bursts of passion and his generosity, even his search for the sublime.

But just how difficult can it be to grow wines? Much depends, of course, on whether the grower is aiming for great wine or simply something adequate for his own table. For fine wines, the job is demanding, as can be seen by following a Burgundian grower from one harvest to the next.

Burgundy has kept the old traditions of densely planted vineyards: from 2,500 to 4,000 vine stalks to the acre, each one requiring the care of the grower. As the average Burgundian estate covers seven to nine acres, this means that there are between some 16,000 and 30,000 plants, each one of which must be pruned, trained, tied, harrowed, buttressed with earth against the winter's frosts and uncovered again in the spring, sprayed and trimmed and finally picked, and the harvest taken to the winery. In a vineyard of seven acres, the pruning alone requires some 400,000 snips of the shears between January and March, in chilly and rainy weather, sometimes under snow.

Pruning is a demanding and tiring task, requiring judgment, a good eye, and a strong, well-trained hand. The snip of the pruning shears influences the future not only of the next harvest, but also that of the following year. Each stalk has its own configuration—even its own personality—that must be taken into consideration. The grower must know his vineyard, and judge it from the harvests it has given in the past. He must take into account variations between one vine plant and the next. The strong one will be given a weightier burden. The weaker ones will have a lightened load, hence a smaller number of "eyes"—the buds from which fruit-bearing shoots will later grow. To grow wine, furthermore, is to exercise the art of predicting the future, knowing that the results will only be apparent in the long run. The winegrower plants for forty years, prunes for two years, and bottles wines which may not be drunk until a decade later.

He must make room in his predictions for numerous contingencies, such as an onslaught of worms, or an epidemic of mildew. He must recruit a team of harvesters a good three months before the harvest—at a time when the date of the harvest is still not known. It is this constant habit of thinking ahead that gives the grower competence as well as circumspection.

When thinking about vineyards, the average wine lover is apt to see only a pretty picture of beautifully lined golden or emerald slopes beneath a crest of forest. From this assortment of greenery, he will seldom distinguish the variety of the grapes, or the particular mode of trimming and training the vine.

In Burgundy, for instance, the finest slopes are shared by three varieties of vines: Gamay and Pinot Noir for the red wines, and Chardonnay for whites. Each one has its characteristics. The plants are not selected in order to make up a pretty mosaic pattern. The choice is the result of the experience of a hundred generations. Chalky soil is beneficent to Pinot. Another type of soil, containing clay, will be just right for the Chardonnay. And yet another, as old as the planet itself, will allow Gamay to give its best. And then, each one of the three rivals will attempt, in its elective soil, to change its personality, to give us a variety of growths whose names are those of villages where each was born. Woe to the grower who tries to ignore these idiosyncrasies. The Pinot Noir from the Côte d'Or, transplanted into the Beaujolais less than fifty miles south, grows moody and lacks spirit. Gamay from the Beaujolais loses its youthfulness when grown in the Côte d'Or and produces mediocre wines. The grower knows all this from instinct and long experience, and can descry that here the land is good for Chardonnay, there for Pinot. In addition, age-old traditions have been consecrated by official decrees, so that transgression is not only heresy, it is also illegal.

NURTURING THE VINE

Before the vine is planted, a graft-bearer must be chosen. Grafting, a serious surgical procedure, has been made necessary by the vineyard's most disastrous pest, a small plant louse, the grape phylloxera, aptly described as *vastatrix* (from Latin *vastare,* to lay waste). The pest came to Europe 100 years ago from the United States, whose native plants have roots too rugged for the phylloxera to damage. The European roots were tender, and nearly all were destroyed by the turn of the century. Today the Old World plants, the source of the inimitable grapes from which great wines issue, are grafted onto American root stock. These are carefully selected to suit the soil, for their health and their vigor, and for compatibility with European vines. The choice is crucial, for it is made for 40 years—the average lifespan of a noble plant. Let's say, then, that the vine has been properly grafted and planted in appropriate soil. It must now be given shape—not just any fancy shape, but one that will be favorable to the grower's work, to treatment of the vine, to proper aeration, to full maturity of grapes.

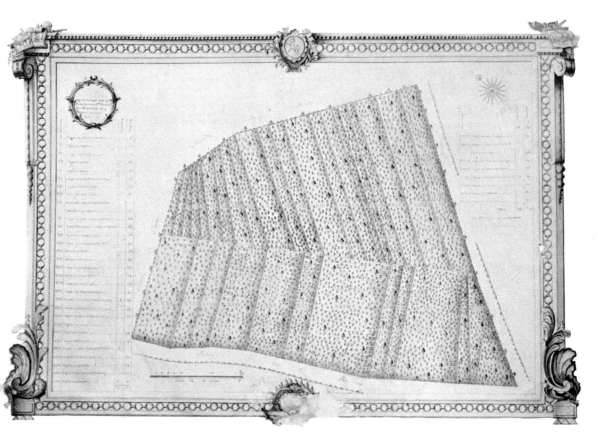

GEOMETRIC PLAN OF A VINEYARD, ancestor of our modern vineyard survey, 17th century. *"L'Imagerie" collection. (Photo Chapman)*

The vine must then grow for at least three years before its grapes may be used to make wine (and even then, the owners of some great vineyards do not consider the grapes of sufficient quality to go into their finest wines).

Then, in the spring of the vine's third year, and every spring thereafter, as the pruning is completed and as the buds begin to swell, sap drips from the wounds made by the shears. "She weeps," the growers say, as they start the next lap in their race against the clock. All of the vine shoots must be trained and tied to lines of wire, and the foot of each vine must be unshod from the earth that was piled up the previous fall to cover it against the cold. Grass and weeds, awakened by the sun, run wild and must be mercilessly rooted out, lest they steal nourishment from the vines.

Throughout the spring—and summer, too—the work goes on. A warning may come from the local oenological station that destructive red spiders have appeared in the vineyards; appropriate insecticides must be sprayed. Rainy weather may favor mildew, a fungus that must be destroyed by spraying with copper sulphate, which must be reapplied if showers wash it away. Oidium, another fungus, may try to destroy the shoots and tendrils, and the grower must guard his vines against worms.

PRUNING VINES THE OLD WAY.
Musée des Arts et Traditions populaires, Paris. (Photo Balure)

VINE ROOT AFTER PRUNING. *(Photo Jean Ribière)*

Opposite page: VINE GROWERS' TOOLS AND THEIR USE. Plate from "L'Encyclopédie," 18th century. *Bibliothèque nationale, Paris. (Photo Chapman)*

As the summer wears on, the grower continues this ceaseless work of weeding, spraying, and pruning, of averting one peril after another, meanwhile keeping a wary eye to the west, fearing the heavy yellowish clouds that signify hail. Should clouds start gathering in force, the growers shoot rockets against them, vainly hoping to disperse them. There is irony in the disproportions between the opposing forces—defenseless man stolidly lighting firecrackers in the face of the oncoming storm.

It is enough just once to see hail pound at a vineyard. Disaster is a matter of minutes. Grapes, leaves, branches all are beaten to the ground by the hail's savage ferocity. And when the clouds disperse, the growers emerge from their homes, dejectedly pace their vineyards, and gather, with a gentleness as touching as it is useless, the debris of grapes that the hail has broken and strewn about the ground.

Such calamity, of course, is not the rule. At the beginning of August, if all goes well, the skin of the new grapes starts turning pink—one here, one there, like stars lit in the sky. Then the grapes become tinged with blue and black. It is a happy time. The growers try to stay away from the vineyard, knowing well that the fragile, budding bunches can easily be harmed. Mildew is no longer a danger—only one major threat remains, that of "gray rot," a tiny fungus which may strike if the humidity rises too high.

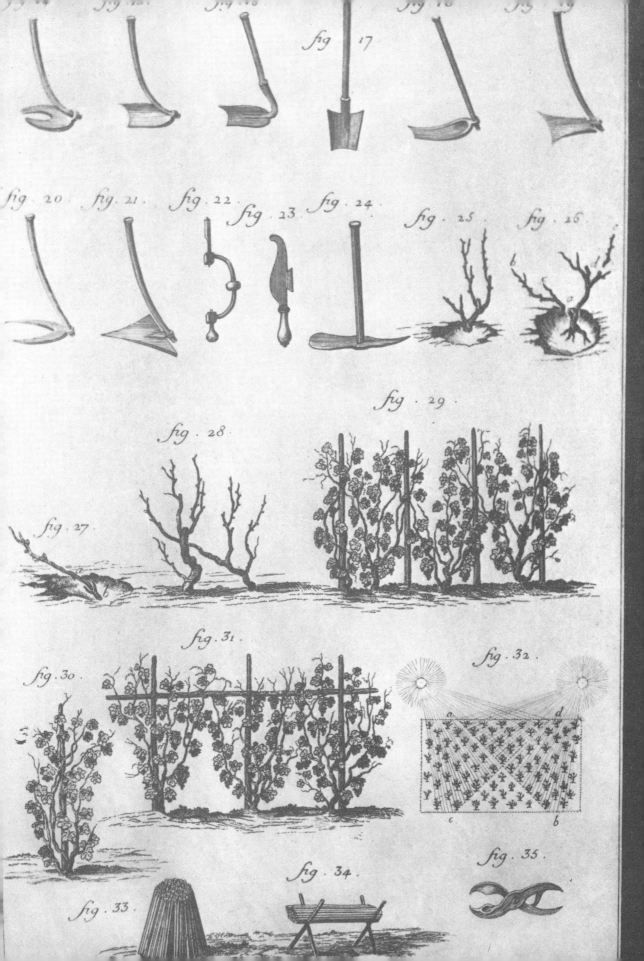

With a song on his lips, the grower readies the cellar where the wine will be made. Hampers, vats, winepresses are cleaned and scrubbed, and barrels stand inspection, are disinfected, and made wine-tight by repeated watering. Pumps, pipes, engines are made ready.

THE HARVEST

As harvest time draws close, oenological experts scout the vineyards, picking a grape here and there, undertaking analyses to determine the degree of maturity. Health bulletins are published in the local press: "The child is in good presentation." Charts show a steady increase of sugar and a decrease of acid content. At a time deemed to be appropriate, the *ban de vendange* (proclamation of the harvest) is decreed: the growers are permitted to start harvesting. The date of the *ban* varies between August 25 and October 20—depending upon whether the summer was sunny and hot, or rainy and cool. Theoretically it is proclaimed 100 days after the full flowering of the slopes.

A form of heroism may be a prerequisite to waiting for the time to harvest—particularly if the growers are asked to wait for the October mists, when cold starts numbing fingers. The fear of seeing the harvest rot exacerbates impatience, and the authorities become insistent that the growers respect the *ban* and wait for the signs of maturity requisite for the quality of the wine—and for the right to label it under the area's appellation. The labor force—hard enough as it is to muster at the end of summer—grows even scarcer with the lateness of the year. When a belated harvest starts, the troops of harvesters are mirthless as they sink into the mist to pick cold grapes covered with heavy dew.

But what a joy when the harvest is early! The days are warm and still long in September, and harvesters, deployed in ranks, forget their weariness to sing the praises of the coming wine.

In the timing of the harvest a major factor is the sugar content of the grapes. In the last days of their maturation, the grapes synthesize sugar corresponding to a daily increase of more than three ounces of sugar per gallon of juice. This

PLEASURES OF THE HARVEST.
Engraving, end of 18th century.
Bomsel collection. (Photo S. Hano)

gain is not apparent to the eye: the grapes already look as ripe as they will ever look, and there is no outward sign of this increasing wealth. The growers may become impatient for the harvest to start—especially if the dangers of hail, or rot (if autumn rains begin too soon), are not over.

French legislators have imposed a minimum sugar content, which varies from region to region, and from one variety of grape to another. And, knowing how easy it is to make a mistake, the National Institute of Appellations of Origin (I.N.A.O.) gives technical assistance and has set up, in all French wine-growing areas, a system of observation of grape maturity. It is under this system that technicians of the I.N.A.O. and official laboratories start taking samples of grapes from selected vineyards about fifteen days before the harvest is expected to begin. These samples are dispatched to the regional laboratory where they are analyzed. After each sampling, every two or three days, the maturity of the grapes is entered on a chart, where it is shown by two curves. One, gradually descending, indicates the acidity. The other, climbing cheerfully, shows the sugar content. Only when the legal sugar content is reached, does the laboratory give the signal for the harvest to start.

Needless to say, some grape varieties mature earlier, others later. The picture can become complex when two or more varieties are mixed into a single wine, as is often the case in the region of Bordeaux and in the south. If one variety cannot be picked separately from another, the choice of the time of the harvest must result from a compromise.

There has been talk in recent years of harvesting machines. Several do exist, notably in California. They are also being studied in Europe, but they require a special pruning of the vine, so that the grape bunches grow approximately on the same level. Because of this, most of the European growers agree that mechanical harvesting will never be suitable to the making of fine wines. But the fact that studies are under way points to the increasing problems of hiring the labor force required for the harvest, especially when it is late.

GRAPEVINE IN THE FALL. Only the grapevine could produce such glorious chromatic variation. *(Photo Legrand)*

Once the harvest begins, a daily minimum must be harvested, so that the vats are not left partly empty. In Beaujolais, for instance, a vat holds about 16,000 pounds of grapes, and should be filled in no more than two days. Since a harvester can pick about 850 pounds of grapes a day, ten harvesters at least are needed. And there must be two men to gather the baskets at the edge of the vineyard, two to transport them into the winery, two cooks to prepare food—in all, some sixteen people.

For an average property having seven to nine acres of vines, the harvest will be around 35,000 pounds, requiring the work of this sixteen-man team for four days. If there is no help to be hired, four growers and their families make up the team, and the schedule must be tightly organized.

TRANSPORTING THE GRAPES

Picking the grapes is only the first step in the harvest. The bunches having been snipped off, the next step is to remove them from the vineyard. Most vineyards are too closely planted to allow a cart to go through, so usually this work is done by hand. On some very steep slopes, as in the Swiss Valais, some regions of the Rhine and the Moselle, and parts of the Beaujolais, small sleds or wheelbarrows filled with baskets of grapes are pulled uphill, either by hand or with a winch and engine. A frequent, and one of the oldest, means of carrying grapes out of the vineyard is the wicker back-basket, often seen in medieval pictures of

HARVEST IN ALSACE AND CHAMPAGNE. *(Photos Chapman and Bernard Jourdes)*

grape harvests. In the Loire Valley, a sort of wheelbarrow with a rubber tire has been designed, with loads of grapes balanced on both sides of a wheel. The most current container in traditional vineyards is a two-handled bucket, still carried on the shoulders of two men—in the same way the warriors of antiquity carried their heroes standing on a shield. Containers are generally made of wood or wicker—seldom of iron, for this metal is wine's first enemy. Traces of iron in the must or unfermented juice can bring about a chemical imbalance, *la casse,* which can ruin the wine. Enough juice can be squeezed out of the grapes while they are being carried in iron containers to cause such damage. Nowadays, plastic and aluminum buckets have made their appearance.

In principle, the smaller the container the better, for too heavy a mass of grapes will start the juice flowing before the harvest reaches the winehouse. This is why Champagne, for instance, has remained faithful to its traditional wicker baskets, and Burgundy to its small buckets. In the Beaujolais, the fruitiness of the wine results in part from the fact that the grapes are undamaged when transported in small buckets. Ideally, the grape should not have suffered a single wound before reaching the crusher. Otherwise there is the threat of oxidation and of premature fermentation.

MAKING THE WINE

The work is by no means finished with the harvest, however, for the grapes must now be made into wine. It is certainly true that fermentation is a natural process: grapes left to themselves with the proper heat and humidity will ferment and transform themselves into wine—wine of a sort at least. The making of a good, drinkable table wine is more complex, and the making of a great wine considerably more so. To understand some of the processes of wine making, it is useful to examine the anatomy of the bunch of grapes.

The skeleton of the bunch is the woody stalk. It contains mostly water, with some tannin, acids, and nitrous and mineral substances. Its most important wine constituent is, by far, tannin. If there is too much tannin, the wine will be harshly astringent (or "puckery"); and if there is too little, the wine will be thin and lacking in body. Tannin, although in lesser quantity, is also supplied by the skins—whose principal contributions are the pigments that will give color to the wine, and fragrant essences that add to its perfume. Still more tannin is contained in the pips—whose number in a grape may range from none to four. The grape itself consists mostly of pulp, which makes up some 85 percent of the weight of the bunch.

The pulp contains all of the elements that will become wine: 75 percent water; 18 percent or more of sugar; acids, and also mineral, nitrous, and pectic substances which play a part in the wine's "body" and freshness, and contribute, together with alcohol and tannin, to its conservation and ageing. Two entirely different types of "musts" (as grape juices destined to make wine are called) can be achieved, depending on whether or not the pips and the skins are kept in

CARRYING TUBS OF GRAPES to the cooperative. *(Photo Chapman)*

contact with the must, which they can alter. Grape stalks are usually rejected, but for red wines the pips and skins remain in the fermenting vat.

It is the skin that contains all of the coloring matter in a bunch of grapes. With the exception of a few rare varieties called *teinturiers* ("dyers"), such as the Jacquez and the Othello plants, whose pulps have color, the juice of any grape, whether the grape is white, red, blue, or black, is always white. This explains the term *blanc de blanc* ("white from white") to designate a white wine made from white grapes, and to differentiate it from a white wine made with red or with black ones *(blanc de noir)*. Depending on the technique of wine making, black grapes can be used to make white wine, rosé, *gris* (gray) or red wine.

Take a few bunches of any current wine-making variety of grape and press them to obtain their juice—or must. For the time being, this juice is inert and, if it is pasteurized by heating and bottled, it will remain a plain fruit juice. However, if the must is left in contact with air, its temperature will rise, and bubbles of carbon dioxide will start coming up to the surface.

This is the beginning of fermentation, caused by yeast, which was long, and wrongly, considered to be an element of the grape itself. It was Pasteur who, around 1860, discovered the nature, origin, and action of the agent responsible for fermentation. He showed that yeast was a micro-organism foreign to the fruit itself, deposited on the grapes, shortly before the harvest, by winds, as well as by birds attracted by the ripening fruit. The yeast can actually be seen, particularly on black grapes, which seem to become covered by a velvety film known as the bloom.

The microscopic, oval-shaped yeast cells have a particular affinity for the sugar found in grapes, which consists not of saccharose (or sucrose, the ordinary beet or cane sugar) but of a mixture in equal parts of glucose and levulose, the two principal forms of fruit sugar. Introduced into the must by the crushing and pressing of the grapes, the yeast cells promptly go to work.

Fermentation is a complex process, still not completely understood. Rather than attempting to describe it in detail, it is sufficient to point out that as soon as yeast cells are introduced into the must, they start consuming the sugar in the juice—and making bubbles of carbon dioxide, which pop at the surface of the liquid. At the same time, the yeast cells release the products of their digestion—alcohol, glycerine, acids and assorted organic matter. While this goes on, there takes place in the yeast colonies a tremendous population explosion, encouraged by whatever oxygen is available.

But then, their gluttony is punished. Alcohol, being by nature a microbicide, kills the yeasts that make it. Fortunately, other yeast cells will take over fermentation, until a certain alcohol concentration is achieved, after which no yeasts can live.

Two and seven-tenths ounces of sugar yield approximately 1 percent of alcohol per gallon of wine. Thus if a gallon of must originally contained 41 ounces of sugar, and if the work of the yeasts has gone smoothly, the wine, once the sugar has been completely transformed into alcohol, will have an alcoholic content of about 15 percent. This complete disappearance of sugar is particularly important for quality red wines, which should never be sweet (with the exception of specially treated wines, such as Port). In a white wine, where a slight sugar content may be desirable, the fermentation must be controlled so that all sugar is not transformed into alcohol. Beginning with a must also containing 41 ounces of sugar per gallon, only about 37 ounces will be given to the the yeast cells. If fermentation is then interrupted, there will remain in the wine four ounces of sugar which would have given another 1.5 percent alcohol: the wine has a slight sweetness.

There are, then, different wine-making methods, depending upon whether white, red or rosé wine is made—dry, sweet or liquorous. These methods differ, notably by the different timing given to two important steps of vinification—crushing and pressing. Crushing, done occasionally by foot, more frequently with crushing machines (essentially toothed or serrated rollers between which the grapes are thrown), breaks the skins and frees the pulp of the grape, at the same time mixing in the yeast cells from the surface of the skin. Pressing, by means of a winepress, extracts juice from the crushed mass of grapes either before or after fermentation.

White Wines

White wines are made more rapidly than red wines, but the procedure may be trickier, mainly because white wines must have a limpid yellow or pale golden-green color, hence should proceed from musts cleared of all solid matter. The clearing of the must makes life precarious for the yeast cells—in addition to eliminating many of them.

Once the harvest is in, the bunches are crushed, with the grapes either left on the stalks or stripped off. Ideally, it would be good to press the grapes without first crushing them, but grapes with unbroken skins would form impermeable masses, making extraction of the juice difficult. Stripping the grapes from stalks has been tried, but it can provoke oxidation and yellowing of the wine.

Whichever method is employed, the important factor is time: wine making must be rapid, in order to limit the contact of the crushed harvest with air, and make sure that the fermenting must does not extract tannin and coloring matter from the solid elements in contact with the juice. Thus small amounts of sulphur dioxide are introduced into the clearing vats into which the must is pumped after pressing to keep the fermentation from starting too soon. It takes six to twelve hours for solid matters to settle down, while the clear juice is gradually pumped off for fermentation.

Sometimes, on the seventh or eighth day, the fermentation seems to stop completely. The unfinished wine hibernates because the temperature has become too low, or because the clearing of the must has eliminated too much of the "food"

PASTEUR'S LABORATORY, showing the apparatus he used in studying the fermentation of wine. *Musée Pasteur, Paris. (Photo Compagnie française d'éditions)*

THE HARVEST. *(Photo Chapman)*

required by the yeast cells. But it can come awake, either spontaneously in the spring, or in response to prodding by the winemaker, who warms up the unfinished wine, gives it air, and keeps the casks filled just to the right level—until, one day, the fermentation is completed, all sugar having been transformed into alcohol.

For making sweet white wines, the procedure is similar—but fermentation must be interrupted before all the sugar has vanished. Fermentation is either interrupted by itself—the alcohol content having become too high for yeasts to survive—or by the introduction of sulphur.

Rosé Wines

Making rosé wine poses a different problem. Wine makers have said that rosé is neither fish nor fowl; some have called it "a monster," others a hybrid. In any event, true rosé is never a mixture of red and white wine.

Up to the 18th century, many ordinary wines were turned out as rosés. In the 19th century, these wines were almost totally eclipsed by reds, and rosé returned into fashion only after World War I.

In some respects, rosé wine is a contradiction, for two aims are pursued at the same time: to obtain the best color, and to have wines that are light, fine, and elegant. These two requirements are in opposition, for a light and diaphanous appearance is usually achieved at the detriment of *bouquet*. Then if, in order to improve the *bouquet,* a larger amount of tannin is given to the wine, the wine will lose some of its graceful transparency. This is a delicate dilemma which can be solved in any one of several ways.

In most of the wine-bearing vines of the Old World, the color pigments are concentrated in the skin of the grape, and are soluble in alcohol but not in cold water. Thus if a bunch of grapes of the Pinot, Gamay, or Merlot variety is pressed, the juice is almost colorless. Only if pressing is very persistent is a light tinge achieved, from squeezing pigments out of the torn grape skins.

When fermentation starts and alcoholic content increases in the vat, the pigments dissolve more readily, and the color of wine gradually becomes darker.

Vin gris, or gray wine, is only slightly pink. The grapes are crushed and pressed, yielding a slightly pinkish must, which is processed as a white wine: sulphur is added to "mute" the fermentation, and solid elements removed 24 or 48 hours later. Then, and only then, is fermentation started. This is the way most Burgundy rosés are made, as well as those of Anjou and of the Touraine.

To achieve rosé wines with deeper color, pigments from the skins must be extracted during fermentation. There are two standard methods. In the first, the mass of crushed grapes is put into a vat, where fermentation starts. Hour after hour, the wine maker examines the color of the juice. When it is deemed to be sufficiently colored, he transfers the grapes into the presses and squeezes out the juice which flows into barrels where fermentation will continue. Most of the time, only juice from the first pressing goes to make rosé—the second and third pressings, too deeply colored, being added to the red wine vats. A second method is to "bleed" full vats by letting wine flow through an opening at the bottom. "Bleeding" is stopped when the color of wine becomes too dark.

The making of sweet rosés follows similar patterns, but the grapes—for instance, Cabernet—are overripe, and the must very rich in sugar. Fermentation is arrested before all the sugar has been turned into alcohol, but too often this requires massive sulphuring, detrimental to the wine.

Rosés, like most white wines, are bottled early, to help them keep their fruitiness. With a very few exceptions, such as those of Tavel, rosé wines do not age well, and should be drunk within a year or two.

Red Wines

The making of red wine is more complex, and techniques vary from region to region. Many important factors are involved.

In southern regions, where it is hot, the juice may have to be cooled off for fermentation to take place; in northern parts, it may have to be heated, particularly if the harvest is a late one. Oenologists have tried to select specific yeasts, in hopes of giving wine certain characteristics: one can, for instance, purchase Beaujolais or Chambertin yeast, in the always vain hope that wine fermented with it will turn up as a Beaujolais or Chambertin. The chief purpose of using selected yeasts, however, is to make sure that fermentation will start rapidly, and will be kept under control until all the sugar has disappeared. To insure this, the fresh must is first sterilized with sulphur, then seeded with new yeast.

Yeast cells do more than just turn sugar into alcohol. They also alter other substances, giving rise to scores of elements which contribute to the wine's character. In fact, yeast cells are not the only substances that can make alcohol and the aromatic elements of wine. Uncrushed grapes placed in a warm place behave like fruits in a fruit loft, and go on living. Even when deprived of oxygen, the cells that make up grapes continue living for some time "anaerobically"—that is, without air—by borrowing from the pulp the elements required for survival.

This phenomenon can be utilized for a particular method of wine making, called "carbonic maceration." It consists of placing the uncrushed grapes in closed vats. The harvest, kept away from oxygen for several days, thus undergoes carbonic maceration, which changes part of its sugar into alcohol without the intervention of yeast cells. The grapes are then crushed, and fermentation is completed with the help of regular yeast cells, in the presence of oxygen. Nevertheless, during the time of maceration, some aromatic substances are born which would not have been created by yeast.

The time factor plays an important role: the duration of the *cuvaison*—fermentation in vats—varies from two or three days to three weeks. The tendency is now to shorten it, in order to produce supple wines that can be drunk without having to age too long. There was a time, in Chateauneuf-du-Pape, for instance, when *cuvaison* went on till Christmas. The result was a tannic and harsh wine, which had to mature for several years to become civilized.

THE HARVEST. A miniature from the *Book of Hours of Guillaume de Bade,* painted in 1647 by Frédéric Brentel. *Bibliothèque nationale, Paris. (Photo Bibliothèque Nationale)*

CARRYING TUBS OF GRAPES at Tain-l'Hermitage, Drôme. *(Photo Chapman)*

HARVESTING GRAPES after the first snow in Alsace. *(United Press Photo)*

The time the juice is left in contact with the grape's solid matters varies from wine to wine, and depends on whether the wine is destined to be kept for many years, or to be drunk while young. The art is to find the best compromise.

Needless to say, it is sometimes necessary to correct nature's imperfections by adding to the must those elements of which there is a shortage—or subtracting those of which there is an excess. For instance, acidity can be too high in cold years and insufficient in hot years; acids can either be neutralized or added. Sugar can also be added. Sugaring is called chaptalization, for Jean Chaptal, who was the first to practice it in France. Some wine makers believe that, if practiced wisely, chaptalization can improve wine, and this is why it is authorized in France when the must does not contain enough natural sugars. In some cases, however, sugaring for the sole purpose of achieving high alcoholic content is harmful, for it upsets the balance of the wine. Chaptalization is altogether illegal in some countries—Italy and Spain, for example. In the United States the practice is often abused, but seldom by the premium wine producers.

Another substance frequently added to the harvest is sulphur dioxide. Sulphur destroys harmful "wild" yeasts, hence helps control fermentation and the temperature in the vats. The burning of sulphur candles in the vat or the barrel goes back at least to the Romans. Today, the burning of sulphur candles or wicks is still used to disinfect wooden receptacles.

During the fermentation of red wines, the solid elements rise to the surface, forming a crust which the French, with good reason, call *le chapeau*. This "hat" has a high concentration of yeast cells, which "work" and make the temperature rise dangerously, so that some cooler must, drawn from below, is pumped over it to cool it off. The "hat" also prevents the must from having contact with the air. Generally it is pushed down with wooden poles, or kept down with a lattice-like wooden frame. Traditionally, *le chapeau* was broken up by men who climbed atop the vat and gleefully jumped in. The "hat" was stiff enough to hold them briefly, before it broke as they jumped up and down, immersing them neck-deep in wine. This practice, called *pigeage*, is now almost extinct.

Watching the temperature, tasting the fermenting wine, examining its color, placing his ear against the vat to listen to the secrets of fermentation, the winemaker waits until he feels it is time for wine to be run out of the fermentation vats. This operation is called tunning.

TUNNING

The wine is far from finished. Tunning, which separates the wine from the solid matter with which it is still mixed in the vat, is but a step, although a step of paramount importance.

For red wines, pressing is part of this operation. The winepress of today is a far cry from what it used to be. In ancient Greece, it was a simple beam of oakwood maintained by various devices above a sort of trough or bucket. Its weight pressed the grapes down; sometimes a heavy stone swung at the end of a lever, adding its weight to that of wood. For the Egyptians, it used to be a simple

THE MUST. Right, a German cellar master checks the alcoholic content with a mustmeter.
(Photos Roebild-Ingrid Autenreith)

sheet of cloth, twisted with the crushed grapes inside. Or—also in ancient Greece, in Rome, and in medieval Europe—it was a set of blunt-edged stakes driven down by mallet to press the grapes.

Then came the screw, one of the great inventions of mankind. At first, it seems, the screw was used to help a beam squeeze down the grape; huge presses of that type, museum pieces to be admired by tourists, can still be seen at Clos-Vougeot in Burgundy and in the German Rhineland. Then came another version, know as the Vaslin press from the Burgundian who originated it, in which the screw stood at the center of a cylindrical, horizontal cage and, moved by manpower, squeezed two circular trays one against the other—grapes being between the two. The Vaslin press is still very much in use today, although the manpower to turn the screw is now replaced by electricity.

The vertical press also has turned to the screw: the result is the traditional and best-known winepress, still widely used, shaped like an upright cylinder of wooden slats, with spaces between them to let the juice escape. The grapes are placed in the cylinder, and heavy wooden boards, forming a circular platform are placed above them. The screw is vertical, fixed at the bottom of the press, with a screw jack to squeeze the platform down. A long lever is grasped by six or eight, even a dozen hands. The men move the lever back and forth or circle around the press, depending upon the gear system. The modern version, once again, eliminates the effort: the press has a hydraulic jack.

Each time a pressful is being squeezed out, all hands assemble to give it a taste: men with their shirtsleeves rolled up, ladies in kerchiefs or young girls in shorts gather around to taste the fruit of the year's labor. The first wine out, called *vin de goutte*—or dripping wine—is best. Then comes the *vin de presse*, squeezed out, harsher and of a deeper red. Both will combine their virtues.

The adolescent wine is then pumped off—or flows—into its temporary home, where fermentation will be completed. In Burgundy, this is the oakwood *pièce* of some 58 U.S. gallons; in Beaujolais, a somewhat smaller *pièce*, 56 gallons; and in Bordeaux, a 57-gallon oak *barrique*. Champagne wines, more exclusive, go to a smaller home, a barrel of 53 gallons. In Portugal (for Port) the cask is called a *pipe*, and holds 151 gallons, the same content as its neighbor across the border, the Sherry *butt*. Chablis, distinguished and light, goes into wood which terms itself a "leaflet," the 36-gallon *feuillette*. In Germany, the cask of wine has *kolossal* dimensions: the traditional retreat for Moselle wines, the *Fuder,* holds one thousand liters (264) gallons). The Rhine wine cask is smaller: the standard is the *Halbstück* (half a piece) and holds 153 gallons. In California, a house of oak is too expensive, and redwood goes to make the 160-gallon puncheon. Elsewhere, the home of wine has followed different traditions. In Turkey and in Soviet Georgia, near the birthplace of wine, the ancient earthenware *karass* is still buried in the earth, but this tradition is vanishing. In Hungary, the famed Tokay goes into wine barrels made in the tiny town of Gönc, which gives its name to 30-gallon casks designed to hold and mix the local nectar. Elsewhere, but not quite everywhere yet, the house for wine is an efficient cement, glasslined vat, standing in rows with its mates. But one should not be too disparaging: if wood is necessary for wines to undergo a slow oxidation and to develop their bouquet, the cement vat is the right place for making wines to be drunk young. Here they are preserved from oxygen, and thus remain unaltered and retain a maximum of fruitiness. Another rule: the longer the maceration of wine with the solid elements of grapes, the longer the wine should mature before it is bottled, so as to have time to transform its tannin into a smooth bouquet.

The process here is beyond understanding. As plant physiologist and chemist Maynard A. Amerine, Professor of Oenology at the University of California, put it (in *Scientific American* magazine, August 1964): "Wine is a chemical symphony composed of ethyl alcohol, several other alcohols, sugars, other carbohydrates, polyphenols, aldehydes, ketones, enzymes, pigments, at least half a dozen vitamins, 15 to 20 minerals, more than 22 organic acids, and other grace notes that have not yet been identified. The number of possible permutations is enormous, and so, of course, are the varieties and qualities of wines." All of these ingredients slowly and subtly work on each other in the vat, and later in the bottle.

When fermentation is completed, the wine contains some 10 to 14 percent of alcohol. Another important constituent is glycerol, a clear, syrupy liquid, which gives wine smoothness. There is from .5 percent to 1 percent glycerol in wine, sometimes as much as 2 percent in sweet liquorous nectars like Sauternes. But

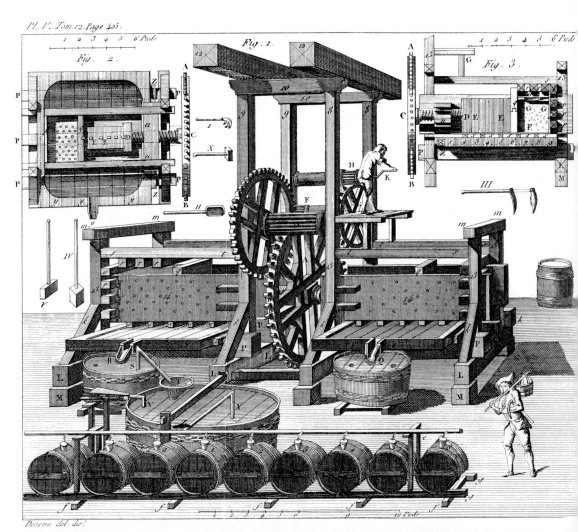

A WINE PRESS IN BEAUJOLAIS,
of the so-called "American style." *(Photo Bernard Jourdes)*

Opposite page: A DOUBLE-ENDED WINE PRESS. Plate from
"L'Encyclopédie", 18th century. *"L'Imagerie" collection.*

wine is infinitely more than that. Pasteur described it as a "living liquid," containing hundreds of other components which, in his words, unite to make wine "the most wholesome of all beverages." There are from .3 to 1.5 percent of acids. Chief among them is tartaric acid; others are malic, citric, tannic or phosphoric, to name a few. Wine also has nearly all known amino acids, which are nitrogen compounds and building blocks of protein, the stuff of life, essential to growth and nutrition. These make up usually less than one part per thousand, and many of them have gone to "feed" yeast cells, so that the average bottle has even less. Wine—red wine particularly—also contains a number of anthocyanins, the principal blue and red pigments of colored flowers and berries, which are, for man, important to night vision. Then there is tannin—not one but a group of substances giving red wine an astringent taste. Vitamins, too, are found in wine: Vitamin A, several of the B complex, Vitamin C, and, more important, Vitamin P, the only one appearing in significant amounts. This vitamin plays a part in strengthening small blood vessels, and its chief source for man is the grape.

Wine contains esters—fragrant extracts resulting from friendly reactions between acids and alcohols. It also contains mineral substances, known as "dry extract," which can be gathered by boiling the wine away. More than 100 substances in wine are known and, according to the well-known French oenologist Professor Jean Ribéreau-Gayon, some 250 chemical compounds have still not been identified.

MATURING IN THE CASK

Wine maturation in the cask is chiefly a process of oxidation, for oxygen seeps through the wooden walls, or reaches wine through the bunghole. "It is oxygen which makes wine," wrote Pasteur. "It is its influence which makes wine age, it is oxygen which modifies the harsh qualities of new wines, and removes from them their bad taste." But ageing is far more than just oxidation. In the bottle, the opposite also takes place: reduction, or the removal of the oxygen component from some of the wine's molecules.

In general, during the first months in the cask, wine becomes less astringent, less deeply colored, and less acid. Its taste becomes refined, while alcoholic content decreases, through evaporation, at the rate of about 1 percent of alcohol every three years for a 50- to 60-gallon cask (the size of the cask is an important factor, as the evaporation takes place through its wood surfaces).

The wine maker must then make sure that the wine is as stable as possible. Ferments and other solid deposits at the bottom of the vat or barrel, known as lees, must be eliminated—either by drawing off the wine without disturbing the lees, or else by filtering. Acidity must be controlled, as well as sugar, color, iron, and other important elements. But there is no routine, no rigid rule or formula, for each vintage is an individual, requiring individual consideration. And the result is always a compromise. The more tests, the more tasting, the more care—the better the compromise.

BOTTLING

Now comes the time for bottling. This operation—which is sometimes performed by the wine lover who buys his wine in casks—should also be approached with circumspection. White wines are bottled six to eight months after the harvest, sweet white wines a year and sometimes more. Beaujolais wines, like other wines distinguished by their fruitiness, must be bottled early to keep their fruit, while red wines such as a Bordeaux from the Médoc must stay in barrels for two years and more. Too early bottling can cause a second fermentation in the bottle, or an excess of heavy deposits of coloring matter or mineral salts. Bottled too late, the wine can have gone flat.

Some wine makers will start bottling only at the new moon, and when the North wind blows—and, who knows?, perhaps they are right. At any rate, absolute cleanliness of all equipment and bottles is required. There should be no abrupt change in temperature from cask to bottle and, as a rule, bottles must be filled with wine almost up to the cork. Wine should not be drunk right after it is bottled: it suffers, in its new glass home, from "bottle sickness," resulting probably from rapid oxygenation during filtering, which dampens its bouquet and sometimes gives it an odd taste. It takes wine one or two months to recover completely—and then, it is just beginning to age.

It may seem natural to the buyer of a great wine that the wine should be limpid, and he may raise his eyebrows at the sight of a speck. This is a wrong approach. It's possible, of course, to filter and even to centrifuge wine so that it becomes crystal clear—but then one has eliminated from the wine the elements that will enable it to gain by ageing. Once more, a compromise must be achieved: if the wine is to be drunk young, if it is destined to travel rather than to remain in a deep underground cellar, it can be cleared of nearly every particle. Nevertheless, the basic principle remains. The more a wine is limpid, the more it is stable, the less its chances of having subtle nuances and distinction. Of course, a vintage wine that has remained in the quiet of a cellar for many years in a horizontal position so that its cork is always kept humid and does not dry out, must first be stood up straight for a few hours, and then slowly decanted, so that deposits are left at the bottom of the bottle.

Therefore, when buying a wine which does not seem limpid enough, don't complain before tasting it. Let the wine speak for itself. Remember that the producer could have resorted to the easiest solution, practically sterilizing the wine to give it brilliance. But then he knows that had he done this, the chances are that this diaphanous but sterile liquid would have lost much of its eloquence.

TREADING GRAPES IN BRITTANY. A Sèvres plate, 1750.
Musée nationale de la céramique, Sèvres. (Photo Ph. Brossé)

and sometimes gives it an odd taste. It takes wine one or two months to recover completely—and then, it is just beginning to age.

It may seem natural to the buyer of a great wine that the wine should be limpid, and he may raise his eyebrows at the sight of a speck. This is a wrong approach. It's possible, of course, to filter and even to centrifuge wine so that it becomes crystal clear—but then one has eliminated from the wine the elements that will enable it to gain by ageing. Once more, a compromise must be achieved: if the wine is to be drunk young, if it is destined to travel rather than to remain in a deep underground cellar, it can be cleared of nearly every particle. Nevertheless, the basic principle remains. The more a wine is limpid, the more it is stable, the less its chances of having subtle nuances and distinction. Of course, a vintage wine that has remained in the quiet of a cellar for many years in a horizontal position so that its cork is always kept humid and does not dry out, must first be stood up straight for a few hours, and then slowly decanted, so that deposits are left at the bottom of the bottle.

Therefore, when buying a wine which does not seem limpid enough, don't complain before tasting it. Let the wine speak for itself. Remember that the producer could have resorted to the easiest solution, practically sterilizing the wine to give it brilliance. But then he knows that had he done this, the chances are that this diaphanous but sterile liquid would have lost much of its eloquence.

TREADING GRAPES IN BRITTANY. A Sèvres plate, 1750.
Musée nationale de la céramique, Sèvres. (Photo Ph. Brossé)

France

BORDEAUX
BURGUNDY
CHAMPAGNE
THE LOIRE VALLEY
RHONE, PROVENCE, AND MIDI
ALSACE
JURANÇON
JURA AND ARBOIS
CORSICA

A "TASTEVIN," OR WINE-TASTING CUP, is a marvel of ingenuity and craft. The buyer's side has large bosses while the vintner's bears finer indentations. *Collection of Mme. De Bruyn. (Photo Chapman)*

Preceding pages: A BURGUNDIAN LANDSCAPE. Its beauties are reflected in the fine wine it produces. *(Photo Legrand)*

BORDEAUX

The crescent-shaped city of Bordeaux is the principal metropolis of the southwest of France. It hugs a curve of the Garonne a dozen miles before that muddy river is joined by the Dordogne to form the wide estuary of the Gironde. And although Bordeaux is some fifty miles from the Atlantic Ocean, it is one of France's most important seaports.

Bordeaux owes its size, importance and prosperity chiefly to wine. All the wines entitled to the label "Bordeaux" come from the *département* (the French administrative subunit) of the Gironde—minus one village (Bazas) and a few small townships. It is not an exaggeration to say that nearly every grown man in Bordeaux has something to do with wine—in addition, of course, to drinking it.

From around Bordeaux, there come the ingredients for a hearty regional cuisine that brings forth the qualities of the wines. There are the *cèpes*, big, fleshy, tasty mushrooms of a variety practically unknown in the United States, which, cooked in olive oil, exhale an unequaled aroma. A few miles east of Bordeaux, in Périgord, there is another type of mushroom, even more celebrated: the delicate black truffle, buried around the roots of oak trees, dug out by pigs or trained truffle dogs. Nearby in the *Landes,* the heaths of Gascony, the speciality is *foie gras,* liver from overfed geese, frequently perfumed with chunks of truffles. The liver may be made into *pâté,* but it is unsurpassed when it is left to simmer for weeks and months in white wine and Cognac, and served intact, as a main course, together with a sweet white Sauternes wine.

There is, of course, the juicy local *entrecôte,* briefly grilled over a fire of vine cuttings, and served with a sauce *bordelaise* made of chopped shallots and red wine. Oysters are "educated," as the French say, off Arcachon, an oceanside resort to the southwest. There are abundant mussels and tiny crabs, both fitting introductions to what is perhaps the region's best gamey *entrée:* woodcock. The woodcock is usually hung down by its feet until ripe—the degree of ripeness being, as with cheese, a matter of personal taste.

These delicacies, or the more proletarian *chabrot* (the dregs of a soup plate to which a glass of red wine has been added) do not seem here to be the privilege of the wealthier *bourgeoisie,* for everyone in the Bordelais—the region around Bordeaux—loves to drink and eat well, and is prepared to make an effort to secure the necessary ingredients. As to the visitor, he is likely to turn into a gourmet if he isn't already one. There are many outstanding restaurants in and near Bordeaux—such as the Splendid Hotel, a favorite of wine merchants and shippers, with a wine cellar housing more than 200 different Bordeaux growths; or Dubern, one flight up and situated above a mouth-watering gourmet speciality shop. Out of town, there are the Lion d'Or in Langon, the Lion d'Or in Bazas, the Hostellerie de la Plaisance in Saint-Emilion, or La Réserve in Pessac.

THE HISTORIC WINES

Statistics irrevocably show that nobody drinks as much wine as the people of the Bordeaux region. And history indicates that nowhere have so many people of so many generations devoted so much time to wine making. Much of the history of France was written with Bordeaux wine and, for that matter, much of British history too. There have been claims—not entirely unsubstantiated—that the British Navy might not have been the powerful force that ruled the seas for centuries had it not been for Bordeaux wines. In 1214, when King John wanted to find out how many ships were moored in Bristol, he had to refer to the wine-standard and order a count of all the ships that could carry at least 80 *tonneaux* of wine, a *tonneau* being the equivalent of 96 cases.

For three centuries, in fact, Bordeaux *was* English, a state of affairs brought about by the marriage in 1152 of Eleanor of Aquitaine, the only child of the last Duc d'Aquitaine, to Henry Plantagenet, Comte d'Anjou, who was crowned King Henry II of England two years later. Aquitaine was the former Roman winegrowing province of Aquitania, of which Bordeaux was the principal city.

ENGLISH BORDEAUX

Along with Eleanor, as a sort of dowry, came the Duchy of Aquitaine and its people, who suddenly became Henry's subjects. The inhabitants of Bordeaux, who were having difficulties selling their wines to other French provinces from which they had been separated by heavy duty barriers, enthusiastically set about teaching their now fellow Englishmen a few things about wine.

They founded the Vintners' Company in London and launched a profitable trade which has never since subsided, not even during the times when Anglo-French relations were strained. One Bordeaux resident, Henry Picard, master of the Vintners' Company, even turned up, in 1335, as Lord Mayor of London. His greatest thrill came when he served Bordeaux wine at Vintners' Hall to five kings at once: Edward III, King of England; King John of France (a "guest" who had been taken prisoner at the battle of Poitiers); David, King of Scotland; Hugh IV, King of Cyprus; and Waldemar, King of Denmark. In 1565, the Vintners' Privilege was granted by charter, giving certain freemen of the company the right to sell wine without an excise license. A very few Vintners' Wine Bars still exist in London, notably Gordon's on Villiers Street and Emberson's in North London, where a good glass of wine can be had cheaply. And the Vintners' Company still gathers occasionally to toast itself:

> *Come, come, let us drink the Vintners' Good Health,*
> *'Tis the Cask, not the Coffer, that holds the true wealth.*

Until the 12th century the wine exported to England from Bordeaux was not much more than the trickle it had been since the Romans started shipping wine to themselves during their occupation of Great Britain, or since the first Christian priests reached the island abandoned by the Romans "to the squalor of unintelligent poverty." The priests did plant vineyards here and there, but the wine was poor, and scarce. Records indicate that in the 6th century, a Norse prince visiting Great Britain was drowned in a barrel of wine, an unfortunate event which indicates nevertheless that some wine trade existed, though it was probably interrupted later, when the Saxons looted the coasts of England.

RICHARD THE LION-HEARTED, king of England.
A popular woodcut from Montbéliard, Doubs. *Bibliothèque nationale, Paris. (Photo S. Hano)*

After King Henry II, the British taste for French wine steadily grew. One of the first serious British wine lovers was Richard the Lion-Hearted, who dwelt and held court in Bordeaux, drank local wine and sang local songs in the *Limousin* dialect of the natives.

John Lackland, Richard's successor, took under his wing the Bordeaux merchants who had begged him to support their trade, and the British love for French wines became passionate. Each year, when the grape harvest started in the Bordeaux region, a British fleet of more than 200 sails left the northern seas so as to reach Bordeaux in the first half of October, when the young wine began to be ready for shipping. Resting in Bordeaux harbor for a month or two, the ships loaded up with new wine to take back to England before the Christmas holidays, when wine sales were liveliest. Most of the wines then sold in England came from Bordeaux. A typical entry by King John's bookkeeper shows the purchase of 267 casks from Aquitaine, as against 54 from Orleans, 8 from Anjou, 26 from Auxerre, and 3 from Germany.

Wine in those days was shipped in and often served right out of barrels, thus it was best drunk in its first youth, before it became oxidized and vinegary.

A WINESKIN
in decorated leather.
(Photo Jean Ribière)

THE PORT OF BORDEAUX.
16th-century engraving.
Bibliothèque nationale, Paris. (Photo Giraudon)

Henry II used to distribute his "old wine" to the poor, and to sell the wine of the previous year at the arrival of the new crop in order to help pay for the latter. The demand was so imperative that in 1255, when King Henry III did not obtain from Gascony the 100 *tonneaux* he required, he instructed that the Bordeaux merchants be called to a meeting and exhorted, "with beautiful and diplomatic words" *(per pulchra et curalia verba),* to sell him more.

In 1307, the royal needs rose to 1,000 *tonneaux*, requested by Edward II for his coronation. Edward felt the shortage of good wine so distressing that he agreed to restore to the *bourgeois* of Saint-Emilion all of the privileges they had lost for rebellious deeds, in return for an annual consignment of 50 *tonneaux* of their good wine.

The citizens of Aquitaine were loyal British subjects, even fought against France, and their wine trade prospered. Only later did other "strange wines"

reach England from foreign lands, Spain and Portugal. In 1381 and 1383, Richard II tasted some wines of Osoye (in Portugal) and of Spain, dubious imports put on the index by Geoffrey Chaucer in the Pardoner's Tale:

> *Now kepe yow fro the whyte and fro the reds,*
> *And namely fro the whyte wine of Lepe,*
> *That is to selle in Fish-strete or in Chepe,*
> *This wyn of Spain crepeth subtilly*
> *In othere wines, growing faste by,*
> *Of which there ryseth such fumositee,*
> *That when a man hath dronken droughtes three,*
> *And weneth that he be at hoom in Chepe,*
> *He is in Spayne, right at the town of Lepe,*
> *Nat at the Rochel, ne at Burdeux toun.*

BORDEAUX UNDER THE FRENCH CROWN

After the Battle of Castillon in 1453 and the end of the Hundred Years' War, Bordeaux returned to the French crown. Its burghers had grown wealthy, and the face of the city changed; fancy houses grew on *rue Neuve* (which still bears the name today) and vineyards spread far outside the city proper, covering almost as much land as they do nowadays.

In France, meanwhile, the use and abuse of wine attracted the attention of legislators. Wines were not yet properly classified, but under Louis XII, at the beginning of the 16th century, wine merchants were. Four categories were described: tavern keepers, who sold wine to be drunk in the tavern; innkeepers, who could serve it with meals; hotel keepers, who were permitted to sell wine only to their boarders; and merchants of wine in jugs, who could retail wine, but could not allow the purchasers to drink it on the spot (to avoid law-breaking by thirsty customers, they sold their wares through a small opening in the door of their store, through which the buyer could introduce his jug—but not himself).

Drinking etiquette was born at the French royal court. In the 16th century, under Henri III, it was a custom to place at the bottom of a glass a crust of roasted bread, and to pass the glass from hand to hand to the guest who was being honored. He emptied the glass and ate the roasted crust, called *toustée* or *tostée*—hence the term to toast, adopted by the English and returned, Anglicized, to the French.

England, meanwhile, started to suffer from an intense shortage of French wines, and began to seek solace on the Iberian peninsula—notably with Port. For political reasons imports of wine from France were heavily taxed, and French statesman Jean Baptiste Colbert was overly optimistic in 1670 when he wrote: "The new duties imposed by England on our wines will not last, as it is difficult, even impossible, that the English do without our wines." In 1703, the Treaty of Methuen dealt a new blow. Signed by the British with Portugal (until then an ally of France) the treaty decreed the imposition of a duty of fifty-five pounds per ton on all French wine imports to England, whereas the duty for Portuguese wines was seven pounds per ton.

This was the time when the hard liquor trade was really born in England, and when an increasing amount of spirits was distilled on the spot. André L. Simon, one of the world's authorities on wine, writes: "For the first time in the history of this land, the country became divided between a large, dazed, gin-drinking majority, and a small, brilliant, wine-drinking, leading minority."

In a Southwark inn there appeared an advertisement:

> *Drunk for 1d.*
> *Dead Drunk for 2d.*
> *Clean straw for nothing*

THE WINE SMUGGLERS

That the British could keep an educated taste in French wines was due largely to the smuggler, who was "liked, encouraged and screened by folks who looked on him as the cheap provider of the water of life and a kind of family doctor." Smugglers provided such wines as Château Margaux (variously referred to in the muddled records of that time as "Margous," "Margoo," or "Château Margou"), Haut-Brion (Obrion, O'Brian, or Houbrion) and other fine, misspelled "clarets." (Claret itself is a corruption of the French "clairet," which referred to a light mixture of red and white wine, sometimes diluted with water.)

A controller of the Board of Customs woefully complained to his superiors that "the boatmen along the coast and the country people favour and assist the smugglers in carrying on their smuggling so much that it is almost impracticable to an officer to get any intelligence from them, and when any poor man happens to bring information to an officer he is afraid of his life, or at least, of being obliged to leave this part of the country." In 1784, it was enacted that a ship should be forfeitable for hovering within four leagues of the coast for the purpose of running goods, even if no liquid cargo were found on board. Smugglers shooting at revenue vessels or officers were condemned to "death as felons, without the benefit of clergy."

But smuggling went on, for it was profitable, and jolly good fun, too. "The

OPALINE GLASS BEAKER, "OCTOBER," with polychrome decoration, late 19th century. *Musée des Arts décoratifs, Paris. (Photo Chapman)*

"KITTY CARELESS IN QUOD."
Cheapside engraving from painting by Rowlandson. *Bibliothèque nationale, Paris. (Photo Cuasse)*

more the fox is hunted the cleverer he becomes and mountains of law may not prevail against winds that are favourable, nights that are dark, and fraudulent folks who are active and resolute," notes an 18th-century observer. Smuggling served a worthy purpose as well, helping people not to succumb to sickly mixtures described in such pamphlets as "The Innkeepers' and Wine Merchants' Guide," the "Vintners' Guide," and the "Bordeaux Wine and Liquor Dealers' Guide," which guaranteed English merchants that they could duplicate the finest wines and liqueurs: "The imitation is in every respect the 'twin' of the real, possessing the same qualities, yielding the same pleasures, performing the same duties, and acquiring and reaching the same end"—though "wine" was sometimes made without the squeezing of a single grape.

Reputable wine dealers, however, managed to keep, sometimes at exorbitant cost, a supply of genuine wines for good customers. There are, notably, many records of the Berry Bros.—who even contributed to the enrichment of literature one day by sending Irish dramatist Richard Brinsley Sheridan a "mulled-up" bill much too promptly to his liking. Sheridan wrote in reply:

You have sent me your Bilberry,
Before it is Dewberry.
This is nothing but a Mulberry,
And I am coming round to kick your Rasberry
Until it is Blackberry
And Blueberry.

By 1860, Gladstone was trying to encourage public consumption of light wines—for the same reasons the Soviets did a century later: to combat alcoholism caused by the habit of drinking cheap hard liquor. He reduced the import duties on wine—a step that finally deterred smuggling, which was threatening to become a British tradition (and, as such, presumably indestructible).

BORDEAUX WINES CONQUER THE NEW WORLD

In the 18th century also, the War of Independence in the American colonies opened a new market for Bordeaux wine. French General Marie Joseph Paul Yves Roch Gilbert du Motier, Marquis de Lafayette, sparked the taste for French wines on the new continent. Lafayette sailed to America from Bordeaux, and an adequate supply of excellent French wines followed him throughout his campaigns. The voyage to the New World showed once more that the hardy Bordeaux wine did not object to a long journey overseas—though the senseless tale that "wines don't travel" persists even today. Most wines do, without being harmed.

At the end of the 19th century, French wine was happily traveling north to the Scandinavian countries and east to the tsar's Russia; across the channel to England, and across the Atlantic to the United States. Of all of the fine table wines imported to the U.S. and Britain today, Bordeaux has remained at the top of the list. Claret—as it still often called today—is drunk immeasurably more in the Anglo-Saxon world than any other imported wine. British and American customers have become so discriminating and demanding that the *Comité interprofessionnel des vins de Bordeaux* (C.I.V.B.), a self-policing organization, employs several *courtiers* (wine brokers and tasters) to taste the wines destined for export.

A FRUITFUL MORNING AT THE C.I.V.B.

Every Tuesday and Friday morning, before breakfast, two or three *courtiers* repair to the spotless white room at the C.I.V.B. in Bordeaux where they find, lined up on a counter, six to a dozen bottles each of white and red wine. Each bottle is anonymous, identified by a number meaningful only to the C.I.V.B. inspector who stands by.

Tucking large white napkins into their shirt collars, the *courtiers* select a wine at random. They pour two or three ounces into a large, classical tulip-shaped Bordeaux glass and examine it against the light. They may comment that the color is "brilliant" and "alive," that the wine has *une jolie robe*—wears a pretty dress. Or that it looks dull, broken or seems "overdressed."

They gently swirl the wine around so that the edges of the glass are moistened up to the rim, increasing the surface that releases its aroma. After inhaling the essence they may find the bouquet "fruity," "voluptuous," "straightforward," or may criticize it as "too austere," "oldish" or "speechless."

These and other expressions of the wine-tasting vocabulary, which may seem meaningless to the casual observer, are not mere affectations uttered by wine snobs. They are perfectly clear to the *courtiers,* whose profession requires them to express elusive qualities of odor and taste.

After such preliminary visual and olfactory examinations, the *courtiers* may already have made up their minds about the wine. Nevertheless they move on to the final test, taking a small sip of wine which they swirl around the tongue and to the back of the mouth, so that the wine can titillate all the available tastebuds. They do not swallow (save, perhaps, at the end, when all wines have been tasted). Seeing them thoughtfully leaning forward, one might think they will let the wine spill out from between their lips, but instead they suck in some air, which breaks up the wine into tiny droplets that splash against the palate and the tongue, and vaporize it into the nasal cavity. The wine is now as eloquent as it will ever be.

Satisfied that they have tasted everything that is to be tasted, the *courtiers* spit the wine out into a long white sink provided for that purpose, rinse their mouths with fresh water, sometimes chew on a bit of bread to help revive the sensitivity of their taste buds, and look at each other.

"Not bad. Perhaps a bit thin," comments the first.

"Thin at first, yes, but it makes a comeback."

"True, it has a comeback—but is it enough of a comeback?"

"How about that milky taste?"

"Disturbing . . ."

"It just won't do."

Almost invariably, they promptly reach the same conclusion and pass on their

COURTIER IN TASTING ROOM.
Faced with unlabeled bottles, he is alone with his conscience
in judging the wine. *(Photos Dorozynski)*

verdict, which is noted down into C.I.V.B. records, where the number of a wine bottle is matched with the name of its grower or with the merchant who wants to export it. The C.I.V.B. does not make detailed comments about the wine, only tersely states that the wine is "accepted" or that it is "insufficient."

WINE BROKERS' PARADISE

Bordeaux abounds in wine brokers—there are several hundred—and it is to the brokers' advantage that wines they sell are kept within trustworthy standards. Brokers, of course, handle all kinds of wine, good or poor; they deal with shippers whose name will appear on the bottle. A shipper's reputation is his most valued asset.

The shippers—such as Barton et Guestier (B & G), Eschenauer, Kresmann, Cordier, Lichine, or Cruse, to name just a few—usually receive brokers in their office, and taste the wine which is for sale. A tag on the broker's flask indicates the origin of the wine, its vintage (if any), alcoholic content, and price. Most of the winehouses have their offices along the Quai des Chartrons, with its decrepit looking row of buildings facing the river across a wide, poorly-paved thoroughfare. Here traffic weaves madly around slow-moving vans and, still, occasional horse carts. Underneath the merchants' offices are their cellars, containing Bordeaux wine worth many a king's ransom.

Frequently merchants blend one wine with another. "The wine trade lives on *coupage,* blending being its only contribution to wine production," writes Alexis Lichine, an American wine lover, grower and merchant with headquarters in Bordeaux. A good blend, of course, can be an improvement, as two wines may have faults that compensate each other. But blending always degrades a great wine by giving it anonymity.

From the shipper wine goes to the wholesaler, by truck, train or ship, to be bottled and sold to retailers. Sometimes a wine is bottled by a shipper; this is the case for most wines exported to the U.S. Shippers also buy wines bottled at the château by the owner. Wine sold under the label *"mis en bouteille au château"* ("bottled at the chateau") usually offers the best guarantee of excellence, but some of the blends of wine from a single region, made by reputable and experienced winehouses, are fine too.

SELECTING A BORDEAUX WINE

The art of loving and knowing wines is not to be learned in books; such knowledge comes with glass in hand. Words can only be props, to help the amateur in the proper direction.

To understand the wines of Bordeaux—or any other wines, for that matter—it is useful to know that the best are not born upon rich, well-fed soil, where grapes might thrive like greenhouse flowers. They come from land where the vine must struggle, from quartz or clay, from dry chalky soil, or sand, or pebbles, thirsting under the August sun. It takes almost a desert to make a wine hearty. Rich earth—like that at the estuary of the Gironde, where much organic matter has been left by the river—yields wines that lack finesse and body, have too much sugar and alcohol.

Because of their heartiness, the wines of Bordeaux are best drunk after having aged longer than wines of any other region—at least five to twelve years after the pressing, depending upon the growth and the vintage year. The vintage year on a bottle is particularly significant, for climatic conditions vary sufficiently from year to year to alter the quality of the wine. (In regions with a fairly constant climate, such as California, indicating the year of a bottle is useful only in that it tells the consumer how old the wine is.) In Bordeaux, the wine is so influenced by weather that the yield of the same parcel can pass from 5,000 barriques (285,000 gallons) one year to 1,000 barriques the next. Similarly, wine of a propitious year, say, 1957, may be infinitely better, and almost infinitely more expensive, than wine of an undistinguished, even though earlier year, say 1956 or 1954. Wines of the Médoc, such as Margaux, Moulis and Saint-Julien, in good years reach excellence five to ten years after the pressing (and they don't start declining until many years later); other Médocs should wait ten to twelve years; Graves and Saint-Emilions, seven to nine. The dry whites are excellent a few weeks after they are bottled, two to four years after the pressing.

The region of Bordeaux boasts some 3,000 *châteaux* and a few *clos* (enclosures). Many of the châteaux are modest farmhouses promoted to castledom only because of the quality of their wines.

THE MAIN BORDEAUX WINE REGIONS

The wine-growing region of Bordeaux could be subdivided into hundreds of smaller areas, each with its characteristic wines but, for the most elementary label-reading purposes, at least five regions should be distinguished: the Médoc, Graves, Saint-Emilion, Entre-Deux-Mers, and Sauternes (including Barsac). It is

CHATEAU LAFITE. *(Photo Jahan, Plaisir de France)*

helpful, also, to know something about classification. The most important one, covering the Médoc and Sauternes, was completed in 1855 by the *Syndicat des Courtiers,* a corporation of wine merchants selected by the Bordeaux Chamber of Commerce to act as experts in the matter. The classification was based on the relative quality of wines over a period of several years, ranking them from *Premier cru,* or first growth, down to *cinquieme cru,* five notches down. Today the classification holds good only as a general rule, with a number of exceptions.

Saint-Emilion wines were classified by a commission appointed by the *Institut national des appellations d'origine* (I.N.A.O.) into two categories, *"premiers grands crus"* ("first great growths") and *"grands crus."* And Graves were classified in 1953 merely by listing a number of outstanding châteaux as *"crus classés,"* either red or white. A total of some 200 châteaux in the Bordeaux region belong to this classified aristocracy, followed by hundreds of *"crus Bourgeois"*—which are by no means ordinary wines, and may be excellent. In addition to these are the *appellations contrôlées,* listing place names of wines corresponding to standards set up with respect to plant variety, maximum yield, and quality control.

The standards of reputable growers are sometimes more stringent than the law itself: the owners of Château Lafite-Rothschild, for instance, have been known to select only the best half of their annual harvest to bear the château's name.

The other half goes to make a more anonymous wine—a plain though excellent Pauillac *appellation contrôlée*.

MÉDOC

Some of the staunchest Bordeaux wines come from the Médoc (a name that comes from the Latin *in medio aquae,* in the middle of the water). It is a triangle of land stretching north from Bordeaux, some fifty miles along the Gironde to the ocean. All of the great wines of the Médoc are red.

Three growths were selected in 1855 as *premiers crus,* the uppermost in constant quality: Château Lafite, Château Latour, and Château Margaux. (Château Haut-Brion, though south of Bordeaux and technically a Graves, was also classified as *premier cru* along with the Médocs in 1855. It is now owned by the family of Douglas Dillon, former U.S. Ambassador to France.)

Château Lafite-Rothschild is a stately and handsome castle with shiny gray slate roofs. Parts of the castle date back to the Middle Ages. Toward the end of the 17th century it belonged to a Monsieur de Ségur who took such great care of the wine that he was known as "the prince of the vineyard." Château Lafite stands on a rolling hill in the Pauillac township, with a vew of the Gironde, in accordance with the ancient Médoc saying that "for the wine to be good, the castle must overlook the water."

Louis XIV spoke of Monsieur de Ségur, as "the richest gentilhomme in my kingdom; for his land produces diamonds and nectar." Louis XV preferred the wines of "Château Laffitte" to any others, and his good friend, Madame de Pompadour, seldom sat through one of her intimate but renowned dinners without a flask of "Laffitte" at hand.

In 1789, Château Lafite became the property of Sieur Pichard of the Bordeaux Town Council, who was elected to the revolutionary *Etats Generaux* and lost his head to the guillotine on the 12th Messidor of Year II (June 30, 1794) in Paris. The castle and vineyards became national property, were later purchased by a Dutchman, who sold them to a French spinster, who sold them to an English banker, Sir Samuel Scott.

German financier Baron James de Rothschild purchased the property in 1858 (for four and half million gold francs) and since then, the castle and vineyards have remained in Rothschild hands, under the name of Château Lafite-Rothschild. Cellars hewn in the rock under Château Lafite are among the oldest in the region, and possess a natural decoration which makes them one of the most beautiful subterranean sights anywhere. There is enough humidity in them to favor the growth of tiny mushrooms *(Racodium cellare)* that weave a thick, velvety moss over the walls and ceilings, forming green, red and gray patches, soft and yielding to the touch, patterned in abstract designs reminiscent of modern works of art. The moss has usefulness as well as beauty: it helps absorb excess humidity

THREE IMPRESSIVE BOTTLES from Bordeaux. *(Photo Chapman)*

that would be harmful to the corks. The temperature down below is constant the year round, and wine unhurriedly matures under the care of a *maître de chai,* or cellar master, responsible for the wine making. His counterpart, the *contre-maître* (foreman) or *chef de culture,* is master of the vineyards, and supervises outdoor work. Most cellars in the Bordeaux area are open to the visitor, whose tour will be the more rewarding if he remembers to address the cellar master not with a plain *Monsieur,* but *Maître.* This traditional form of address is used today only for cellar masters, attorneys, fencing instructors, and great painters, composers, and conductors.

Château Lafite and other "clarets" ferment for eight to twelve days, depending upon the weather conditions. Throughout the fermentation, the wine remains mixed with the seeds and skins of the grape, which give it the red pigment absent from white wines. After the fermentation, the wine stays in the vats for as long as three weeks to allow a first settling and elimination of the lees, or sediment, that sinks to the bottom. It is then pumped into the traditional 57-gallon oakwood *barriques* where it is periodically clarified with egg-whites, then "racked," or drawn out, into another cask. Drawing out is a solemn, silent operation performed by candlelight, which is said to be better than electric lighting in helping to detect, in the stream flowing from the cask, a deepening redness which signals that the level of the wine is down to the lees. After several clearings and "ullings" (refilling the casks through the bunghole), 10 to 15 percent of the harvest has been eliminated, but the wine is limpid.

Château Lafite boasts the most impressive wine collection anywhere, *la réserve du propriétaire.* In the deepest vault under the hills, remote from the noise, hustle and vibrations of everyday life, lies the owner's personal stock of 80,000 bottles. The oldest date back to the French Revolution. (This reserve survived the German occupation during World War II because Hermann Goering had set Château Lafite aside for his own use after the victory.)

A routine has been established to open and recork each bottle every twenty years so that the wine does not take the odor and taste of old cork, or escape through a damaged cork. On a rare occasion, a venerable bottle is carefully extracted and brought to light to end a life spent in dark seclusion, and carefully decanted into a crystal home. The best of the reserve tasted in recent years was served at a dinner in 1959. It was the centenary 1859 vintage, which had kept exceptional bouquet, body, and strength.

Near Lafite stands Château Latour ("the tower") also in Pauillac. The name dates from the Middle Ages, when fortifications and towers had been built against seagoing pirates who sailed up the Gironde. Latour was an advanced post in many skirmishes—the most famous during the Hundred Years' War, when Bertrand Du Guesclin, Constable of France, stormed it and took it from the English. Château Latour has the most full-bodied of the great Bordeaux wines.

Château Margaux, the "king of the Médoc," in its best years is without a peer, providing a wine that is unmatched for delicacy and finesse. It is, however, hypersensitive to weather, and in years of misfortune can approach mediocrity.

CHATEAU HAUT-BRION. 19th-century engraving.
"L'Imagerie" collection. *(Photo Chapman)*

"I Am Sheep"

Just across from Château Lafite, Château Mouton-Rothschild is the property of another branch of this family, the descendants of Baron Nathaniel de Rothschild. Purchased in 1853, Château Mouton-Rothschild was classified in 1855 as a "second great growth". It was a prestigious enough title, but disdained by its owners, who stood aloof from the rating and proclaimed their own motto: *"Premier ne puis, second ne daigne, mouton suis"* (which can roughly be translated as "I can't be first, don't condescend to be second, I am sheep"). Mouton is French for sheep, but the château's crest is a ram. The wine certainly deserves to be rated among the first growths; it is one of the most obvious examples demonstrating the obsolescence of the 1855 classification.

At any rate, do not overlook the "second great growths" of the Médoc; many are worthy of intense attention. In good years, they can be better than some of the first growths. And many outstanding wines can be found among the third, fourth, and fifth growths. Examples of fine château wines among the lower growths include: Second growth—Rauzan-Segla, Lascombes, Leoville-Lascases, Pichon-Longueville-Comtesse de Lalande, Cos-d'Estournel, Ducru-Beaucaillou, and Brane-Cantenac; third growth—Palmer, Grand-La-Lagune; fourth growth—Prieuré-Lichine; fifth growth—Lynch-Bages.

GRAVES

The region of Graves—meaning gravel—lies west and south of Bordeaux along the left bank of the Garonne river. The Graves wines were classified in 1953, and the *appellation* Graves Supérieur is restricted to the white wines, though there are excellent, and often more expensive, *appellations contrôlées* reds.

The best red Graves have a likeness to Médoc wines—and, as we have already noted, Château Haut-Brion, the outstanding red Graves, is actually classified with the Médocs in the 1855 classification, but not as an *appellation contrôlée*. Château Haut-Brion also produces a dry white wine, one of the best in Bordeaux, but little known because it is extremely scarce. This wine is made with the same plants that produce the sweet Sauternes wines of Château d'Yquem—which goes to show how much difference the soil can make. The soil of Haut-Brion is credited with producing good wines even in particularly poor years: its gravel, sometimes as much as 50 feet deep, provides a particularly efficient draining system. Some of the vintages of Château Haut-Brion—such as the 1928—reach their peak only more than thirty years after the harvest.

There is a story, apocryphal, about Château La Mission Haut-Brion, suggesting it as the site of a miracle of the Lord. The Eternal Father, it is said, had dispatched St. Vincent, patron saint of vintners, to explore the wine-growing regions of France. When Vincent visited La Mission Haut-Brion, he was so well received by its monks, and so liked their wine, that he could not remember his way back to paradise. In a moment of anger, the Good Lord turned him into stone, in the very cellar where he had sinned. Today visitors at La Mission can still admire the handsome statue of St. Vincent, with its eternal smile.

Other Graves red wines worthy of note are Château Pape-Clément (founded in 1300 by the Archbishop of Bordeaux, who became Pope Clément V a few

CHATEAU AUSONE, Saint-Emilion, Gironde.
A typical château of the Bordeaux region. *(Photo René-Jacques)*

years later), Château Smith Haut-Brion, Château Haut-Bailly, Château Smith-Haut-Lafite, and Château Carbonnieux.

Château Carbonnieux is also outstanding as a *white* Graves. It is one of the oldest wine-growing domains in the Gironde, once the property of the Benedictines of Saint-Croix, who exported their wines the world over, even to the land of Moslems, who are forbidden alcoholic beverages. The legend has it that the wine was shipped to the favorite mistress of a Turkish sultan, after being baptised "Mineral Water of Carbonnieux." Dry and fine, with a vigorous fruity bouquet, the excellent Carbonnieux white wines are somewhat reminiscent of Rhine wines.

SAINT-EMILION

The beautiful, colorful little village of Saint-Emilion is comprised of convents, chapels, historical monuments, and old stones. Underneath the village there is a church carved from the rock, and a cavern from which St. Emilion, the hermit, used to contemplate the world in the Dordogne valley below. Next to the cave there is a small well, into which eligible girls throw their hairpins: if the hairpins form a cross at the bottom, they can expect to be married within the year.

Formerly a pilgrim's way-stop, Saint-Emilion is surrounded by communities bearing the names of local or itinerant saints: Saint-Christophe-des-Bardes, Saint-Laurent-des-Combes, Saint-Sulpice-de-Faleyrens, Saint-Etienne-de-Lisse, Saint-Pey-d'Armans, and Saint-Hippolyte. Together with the township of Vignonet, these district produce wines that can be labeled Saint-Emilion.

The red wines of Saint-Emilion, growing on some 17,000 acres of land, are a bit harsh, sometimes hard and slightly bitter when young, but the best of them acquire great distinction with age. They are, by and large, more full-bodied than Médocs and Graves, a quality which has earned them the name, "the Burgundies of the Gironde." Two names stand out: Château Ausone and Château Cheval-Blanc.

Château Ausone vineyards are said to have been planted in the fourth century A.D. by the poet Ausonius, tutor to Emperor Gratian. The generous "wine of the poet" can stand comparison with the great Mêdocs. Château Cheval-Blanc takes its name from a white horse purchased there by King Henry III in the 16th century, a horse he later rode into battle. This domain produces a full-bodied wine, yet gentle to the palate. Other worthy Saint-Emilion neighbors are Château Beauséjour, Château Belair, Château Canon, Château Figeac, Château Pavie, Château La Gaffelière-Naudes, and Clos Fourtet.

The nearby township of Pomerol—with only 1,500 acres of vineyard—is the smallest district of Bordeaux and, some say, the most distinctive one. Even the ordinary *appellation contrôlée* Pomerol has a deep bouquet of its own and a velvety, smooth, almost unctuous quality. Three neighboring communities, envious of the Pomerol fame, asked to bear the Pomerol *appellation* and the matter had to be settled in court: only a small portion of Libourne was given the honor, while Lalande was bestowed the *appellation* of Lalande Pomerol, and Nérac got nowhere. The judgment, confirmed in district court, reflects the individuality of

Pomerol wines—which have never been classified. Outstanding among the Pomerol growths is Château Petrus, whose reputation even its neighbors will not dispute. The neighbors themselves, however, are worthy of consideration: Château l'Evangile, Château Lafleur, Château Certan, Château Latour, and Clos des Templiers.

ENTRE-DEUX-MERS

Between the Dordogne and Garonne rivers a picturesque tongue of land, the Entre-deux-Mers, produces some excellent dry white wines. They accompany *fruits de mer* (shellfish and crustaceans) so well that they have given rise to a saying: *"Entre-deux-Huitres—Entre-deux-Mers"* ("Between two oysters—Between two seas").

Along the Garonne, southwest of Bordeaux, the Entre-deux-Mers becomes the Première Côtes de Bordeaux. This region offers honorable red wines and some well-appreciated white wines. Among them are Loupiac, Sainte Croix du Mont, Sainte Foy Bordeaux and Graves de Vayres.

The *"palus"* (swamp) part of the region produces only wines that lack body and finesse.

SAUTERNES

Seldom is nature as close to man as it is in the rolling hills of the Sauternes region, some 25 miles southeast of Bordeaux on the left bank of the Garonne.

If you were served ripe Sauternes grapes for dessert, you would refuse, politely but firmly. Each grape is withered, mottled and sickly. It is covered with a whitish mold which, upon close inspection, turns out to be a microscopic, mushroom-like fungus. To put it bluntly, it looks like the grapes are rotten.

Indeed, agrees the winegrower, they are—and must be. But this is an aristocratic mold, *pourriture noble,* the work of *Botrytis cinerea,* a tiny mushroom that concentrates in the grape the sweetness and the bouquet which you will find only in the liquorous golden wine with the authentic Sauternes label, and in the few German *Trockenbeerenauslese* wines, which are not even produced every year.

Wine making here is particularly tricky, because *Botrytis,* which occurs spontaneously in Sauternes and spreads with gentle encouragement from the moist, misty mornings that shroud its slopes, is capricious. More often than not, it settles only on a small portion of a bunch of grapes, properly withers and wrinkles it, while the rest of the grapes hang on, disappointingly plump and healthy.

As a result, the harvests in the region are painstaking and slow. The harvester snips off with scissors only the moldy part of the bunch, leaving the rest until it is shrunken, in turn, by the noble mold. A week or so later, the harvesters will snip the vine once more, and again, a third time, sometimes even a fourth, before all of the harvest is in. On at least one occasion, the mold was so slow in forming that one château did not finish its harvest until Christmas.

The dry, chalky soil that prevents the grape from growing too plump with too much water, the withering and sweetening process accomplished by *Botrytis cinerea,* the subsequent secretions of the "noble mold," its interaction with the yeasts of fermentation—all are natural links along a chain which must be put together patiently by man, who helps along a miracle of wine making he does not quite understand.

Of all Sauternes wines, the most prestigious comes from Château Yquem, boasting the title of *"grand premier cru de Sauternes,"* or "great first growth." Let us follow the grape (mainly of the Sauvignon and Sémillon varieties) on its way from the plant to the bottle on this 360-acre domain owned by the Marquis de Lur-Saluces.

Grape gathering starts comparatively late, usually around mid-October, depending upon the heat of the summer, the rains, the ripeness of the grape and, of course, the degree of contamination by the *pourriture noble*. The harvesters—old shrunken women in black woolen dress and shawl, veterans of the "grape-by-grape" picking method, or young girls wearing more colorful garb—are driven in every morning from the surrounding *Landes*. One for each row of vine, they chat rapidly and progress slowly, snipping off with pruning scissors only those grapes marked with the noble mold. Slowly also, *les bastes*—the round, wooden buckets waiting at row's end—become filled.

Followed by the buzzing of banqueting bees and the heady scent of beginning fermentation, the buckets ride up the hill to the château in flat carts pulled by heavy horses whose pace is enlivened by the lightness of their meager load. ("Tractors? Why should we have tractors?" wonders one old vinetender. "Unless you find a tractor that produces manure!") Then, in a white cloud of Botrytician dust, the buckets are emptied into a mangle where skins are broken and stalks thrown aside.

Hydraulic presses go to work, and the thick "must" flows into cement vats below the floor. The harvest of the day is mixed in these vats to achieve a uniformity, for grapes picked in the early morning and still covered with dew would make a lighter wine, while those collected in the afternoon would yield an essence too heavy, too sweet, and too strong. In the evening, the "must" flows into

THE CELLAR MASTER TASTING HIS WINE. Under his watchful eye, the wine slowly matures to greatness. *(Photo Jahan, Plaisir de France)*

oakwood casks, made new every year for the year's harvest to profit from the subtle interplay between fresh wood and young wine.

Meet Le Maître De Chai

Once in the cask, the wine enters the realm of the *maître de chai,* who will nurture and nurse it for three years, until bottling time. The *maître,* here, has undeniable authority. He is not the rather stout, red-faced man one likes to imagine nonchalantly leaning against a cask in the dimly lit cellar of an elegant château, dipping a syringe-like glass "thief" into the wine to fill a long-stemmed crystal glass. He has heavy responsibilities, that of the well-being of three to four hundred *tonneaux* of Château Yquem: the two year-old wine, soon to be bottled, last year's growth, still unpredictable, and the newborn, the most tempestuous of them all. *Maître* Bureau (Bureaus have been the cellar masters at Yquem for some 400 years) works in close cooperation with the Marquis de Lur-Saluces, the owner, and with an agronomist.

Now he must see to the new wine. In Sauternes, fermentation could bring it to the highest natural alcoholic content anywhere—upward of 18, sometimes as high as 25 percent. But fermentation must be stopped much earlier, before all of the sugar has been transformed into alcohol. Sometimes this happens by itself as *Botrytis cinerea* turns into botriticine, an antibiotic, which counteracts the yeasts. The ideal outcome of this micro-organic struggle is, then, the achievement after some three weeks of fermentation of a spontaneous alcohol content of 14 to 15 percent, balanced with a sugar content of 4 to 6 percent.

But usually this does not "just happen." One year, the ferment may be particularly pugnacious; another year, botriticine unusually strong. The tug of war

GRAPES SHOWING THE "NOBLE ROT"
(Botrytis cinerea)
at Château Yquem.
(Photo Bernard Jourdes)

Opposite page:
THE CELLAR MASTER
at Château Yquem.
(Photo Bernard Jourdes)

becomes unfair and the *maître de chai* must interfere—to slow down, for instance, tempestuous fermentation by throwing in some sulphur, which attacks the ferment.

Maître Bureau is not a microbiologist but a true artisan, relying upon his long-trained palate, eye and nose, and a judgment which seems, somehow, inherited from a long lineage of *maîtres* before him. But he is also versed in modern wine-making techniques, which he applies to double-check his "feeling" for the right thing to do, like a physician who confirms his diagnosis with the appropriate laboratory test.

Then comes a day—a cold and wintry day when nature's forces are at rest—when the wine is limpid enough, mature and quiet, impregnated with a delicate yet stable enough bouquet. It is bottling time.

After bottling, wines of Sauternes, and notably Château Yquem, are capable of longevity man himself can envy, for it is long as well as gracious. Of a light amber color when it leaves the cask, the wine of Château Yquem takes on with age the remarkable depth of old gold. Most wines, even the *grands crus* of the Bordeaux region, should be savored within ten, twenty, seldom thirty years after bottling. But coming from Yquem many a bottle of a good year has grown to be a hundred to yield incomparable delight. Past that age (and earlier for less fortunate vintage years) it becomes highly unpredictable, for it is subject to *madérisation*—a change toward the taste and color of Madeira wines, unsuitable for so distinguished a Bordeaux.

In spite of modernized equipment, wine making in the great château must remain traditional, unyielding to the modern trend toward mass production, undaunted by the hustle to do more in less time and with less effort. To hold its own amid high competition, a wine that is costly to buy because it is expensive to produce must be without reproach (or at the most with only charming vices).

A visit to Sauternes is not complete without a taste of some of the other great growths, either from the Sauternes township itself, or from neighboring Bommes, Preignac, Barsac, or Fargues. Château Latour Blanche produces a wine with a noticeable flavor of muscat. Château Rayne-Vigneau approaches in excellence Château Yquem; and it is a domain which has in its soil an abundance of semi-precious stones—onyx, agate, white sapphire and such, a collection of which can be seen there. Finally, Clos Haut-Peyraguey, Château Lafaurie-Peyraguey, Château Coutet, and Château Rieussec have enviable reputations.

The *appellations contrôlées* here may lead to some confusion. Barsac, for instance, is a parish of the Gironde next to Sauternes, and pretty much wedded to it—but without giving up its name. The vineyards of Barsac produce a lot of white wine, all sweet. But many of these wines are sold as plain Sauternes, others as Barsac, and some as Haut-Barsac (that part of Barsac further south which is geographically more elevated), or as a combination of the three. A true Barsac, such as Château Baulac, then goes under the hybrid label of Château Baulac, Haut-Barsac, Sauternes.

Be that as it may, we must agree with Michel Eyquem de Montaigne that the wines of Sauternes are "the quintessence of the grape and of the sun," and follow his advice to "accept gratefully what nature has given us . . . One offends the great and almighty Giver by refusing his gift, by altering or disfiguring it."

PRUNING HOOK WITH IVORY HANDLE, 17th century
Musée des Arts décoratifs, Paris. (Photo Chapman)

BURGUNDY

To the average Frenchman, Burgundy is the land of jolly and plentiful meals, drinking songs and wine-drinking fraternities, snails, and deep-vaulted cellars silently guarding their priceless liquid assets. To the wine lover, of whatever nationality, it conjures up the image of an old bottle, dusty or mossy, ready to display its ruby-colored liquid from which wafts a mysterious and haunting perfume, or the image of a dry white nectar unequaled on this planet.

The very names of the great Burgundian growths are standards bowed to the world over: Chambertin, Bonnes Mares, Romanée-Conti, Corton Charlemagne, Montrachet, and others, enough for the fulfillment of any wine lover's dreams. There is wine for every season, every moment, every mood, and every meal.

The land itself is pleasant and varied, with lakes and pastures, valleys and forests. Smiling and orderly, it is dotted with peaceful villages and famous castles and crossed by gentle rivers. Like the landscape itself, the people are reasonable and deliberate, yet with enormous zest for life.

The "best wines of Christendom" are many and varied, despite the comparatively small acreage involved. The most famous come from the Côte d'Or, the Golden Slope, a ridge of hills extending some thirty-five miles from just south of Dijon towards Lyon. This is divided into two sections—the Côte de Nuits to the north and the Côte de Beaune to the south. South of the Côte d'Or are the southern Burgundy districts: the Mâconnais, the Côte Chalonnaise and Beaujolais; and to the northwest, toward Paris, the small but justly renowned vineyards of Chablis.

BURGUNDY WINE LABELS

The system of labeling wines in Burgundy may seem hopelessly complex and confusing (there being over one hundred varieties). There are, however, a few clear-cut and simple categories, easy to recognize.

The most ordinary wines are sold simply under a generic regional label: the *appellation contrôlée* of *Bourgogne* (Burgundy), or Bourgogne Supérieur. These are produced under strict governmental control, from prescribed plants and in limited amounts, and can be good, although they are not greatly distinguished. In terms of quantity they are the most abundant. Wines of higher quality are given the name of one of the districts in Burgundy, such as Côte de Nuits or Beaujolais.

A notch above are wines entitled to carry the name of the producing township. Examples of these would be Gevrey-Chambertin, Pommard, or, in the Beaujolais, Fleurie or Moulin-à-Vent. Wines thus labeled are of more distinction than simple Burgundy, although not yet on the highest rungs of the ladder. Yet another step up are wines labeled with township and vineyard. Gevrey-Chambertin Clos St. Jacques is an example. This indicates the wine not only comes from Gevrey-Chambertin (and meets the standards set for wines that may bear that name) but from the particular vineyard of Clos St. Jacques and has met even higher standards. At the very top are the great vineyards themselves: Chambertin, Corton, Montrachet—wines whose names are pronounced by connoisseurs with a note of reverence in their voice.

BEAUJOLAIS

The traveler heading north through Lyon can see from the outskirts of that industrial city the rounded slopes of the Beaujolais, the producer of some of the world's most amiable and popular wines. Vines abound, flourishing from the foot of slopes to a height of 1,800 feet or so and are to be found almost everywhere, except along the banks of rivers where the humidity is too high.

The district is overwhelmingly dominated by one variety of vine, the Gamay, also called Gamay Beaujolais. Grown in tight ranks, one plant barely more than a yard from the next, the Gamay vine looks like a five-branched candelabra, a shape achieved by pruning, and intended to give it maximum exposure to the sun. Some 4,000 plants per acre are crowded into these vineyards.

As for the wine, it is particularly pleasant to drink in its native habitat, for Beaujolais is nothing if not hospitable country. Each village has a cellar set up for visitors, where the growers set out their bottles for visitors to taste, attracting in spring and summer lines of southbound tourists.

There are nine "noble growths" of Beaujolais. These, going from north to south (but not in a straight line), are Saint-Amour, Juliénas, Chénas, Moulin-à-Vent, Fleurie, Chiroubles, Morgon, Brouilly, and Côte de Brouilly. Each has its own qualities and characteristics, but nonetheless the wines vary greatly from year to year, grower to grower, and vineyard to vineyard. A prize offered two decades ago for the growth that would be judged best for three years in a row has yet to be awarded. Ranking slightly below these noble growths are wines that call themselves "Beaujolais Villages," sometimes also carrying the name of the township where they are produced, such as Villié-Morgon, Saint-Lager, Montmelas, or Denicé. Other wines from the district are labeled Beaujolais Supérieur or simply Beaujolais. These are the more ordinary, everyday wines, usually served in a decanter or in the classical *pot*, a bottle with a wide neck and a thick bottom.

IN OFFERING YOU A TASTE, the vintner pours out a little of his heart.
(Photo Legrand)

HARVEST AT
"CLOCHEMERLE"
(Photo Bernard Jourdes)

Most of the wines are red, although in recent years some growers have begun producing increasing amounts of rosé and white wines, both of which are also legally entitled to the name of Beaujolais. White Beaujolais can be made from either of two grapes. Ordinary wines are made from the Aligoté grape, but the better ones, having a similarity to the white wines of the Mâconnais district to the north, come from the Chardonnay.

Beaujolais wine, some 20 million gallons of it a year, is made from 60,000 acres of vineyards by some 6,000 to 8,000 growers, most of them working under a type of contract, peculiar to the region, called *vigneronnage*. It is a form of crop-sharing, whereby the grower takes half of the yield, and the landowner the other half. The owner supplies the wine-making machinery and whatever products may be required for the vineyards throughout the year, while the grower furnishes the viticultural equipment, labor, and any animal or mechanical power that may be needed. To Burgundians, this is a most fortunate association between labor and capital.

Beaujolais is among those wine-growing regions most closely clinging to tradition—not for tradition's sake, but because here tradition is an essential requirement. Limited in the space of his dozen or so acres, restricted by the ups and downs of the slopes, the grower has largely remained faithful to the horse as a mode of propulsion and transportation, and continues to do most of the work by hand—not for the love of manual labor, but of necessity.

The harvesting techniques have changed but little in several hundred years—

and this has helped Beaujolais wine keep its individuality. The notion of *terroir* (a parcel of land) and of *cuvée* (the vatful of wine issued from that parcel) bespeaks diversity. The fact that grapes are carried on man's back itself helps quality: the baskets must be relatively small, and grapes reach the winehouse without having been crushed and their pulp exposed to air, which would tend to start untimely fermentation.

To define Beaujolais wine with precision would be to define charm, to explain harmony, to measure with precision a degree of pleasure. This is beyond our means. Rather, let's say that Beaujolais is simple, easy-going wine. Its drinking brings to the mind happy and reassuring images. Its flavor can be classified in a gamut of floral or fruity scents. It seems to hold within itself the warmheartedness of the people who made it. It is thirst-quenching rather than assertive. It has more spirit than body, more grace than shape, more elegance than pretension. It is fresh, sprightly, roguish, malicious and coquettish.

Beaujolas is one of the world's finest table wines. All the red growths lend themselves well to be drunk cool—at cellar rather than room temperature—and are suited to a wide variety of food. They are wines made to be drunk young—in January following the harvest, and throughout that year, until the "new wine" is ready and bottled. This is not to say that Beaujolais doesn't age well; it can, but then it acquires completely different qualities—distinction if the growth is good, and a new bouquet.

THE MACONNAIS

The regions of Beaujolais and Mâconnais are so closely interwoven that one district, La Chapelle de Guinchay, 10 miles south of the town of Mâcon, can be called either Mâconnais or Beaujolais. The difference between the two regions, however, is noticeable. Beaujolais has the relief of ancient granitic mountains, with rounded peaks and gentle slopes, while the Mâconnais has a chalky geological past, with horizontal crests, broken by abrupt ravines. The soil itself is different: the pinkish hue of the granitic slopes gives way to ochre, chalky soil.

At Leynes, last village in the Beaujolais and first in the Mâconnais, there is no longer any doubt. The visitor finds himself in a sort of crater covered with vineyards. Not far away, the houses of the town of Mâcon are silhouetted in the sky; below are the vineyards of Pouilly-Fuissé—a fitting prelude to this region of white wines, the kingdom of the Pinot Chardonnay, the same plant that is used to make the best white wines of Burgundy, and all Champagnes.

The vineyards look more orderly, and more severe. No longer are the growers' houses gaily scattered among them. They cluster near the scarce watering points, crowding around church steeples. Opening on the large valley of the Saône river, the panorama of the Mâconnais is organized like a grand menu: to the right, Bresse, a gourmet country famed for its chickens; to the left, a wine list worthy of the choosiest wine lover.

A CONNOISSEUR of Beaujolais. *(Photo Brassaï)*

A wine road has been mapped for travelers. It is like a snake, gray and meandering from village to village, between vines pruned so that they look as if they were curtsying, with rounded arms. The villages are Viré, Clessé, Lugny, Azé, Verzé. Then Chardonnay, Saint-Gengous de Soissé, and Igé. The road then skirts around Mâcon, and reaches Vergisson, Solutré, Fuissé, and Chaintré.

Aside from producing some Beaujolais in its southern part, the district yields wine labeled under the Mâcon *appellation contrôlée,* red, rosé or white. The red wine is pleasant and sound, usually less fruity than a good Beaujolais. All these wines can be tasted at the Maison Mâconnaise des Vins, solidly planted at the gates of Mâcon.

The white wines can have more distinction—particularly those grown in the district of Pouilly-Fuissé, Pouilly-Loché, Pouilly-Vinzelles, and Pouilly-Solutré. Pouilly-Fuissé is a worthy companion of seafood of all sorts. It is a pale, golden-yellow, vigorous wine; it can be drunk in its prime youth and gains little if at all from being aged for more than five years. Pouilly-Vinzelles is usually lighter and shorter-lived, and Pouilly-Solutré more delicate. White wine labeled Mâcon possesses, though to a lesser degree and with less distinction, qualities similar to those of the Pouilly wines.

During the reign of Louis XIV, the Sun King, Mâcon wines knew a period of disfavor, which prompted one of the growers, Claude Brosse, to load a few barrels onto an oxcart and drive it some 300 miles to the court in Versailles. In a time when French roads were infested with brigands, it took some courage to undertake the lonely trip; nevertheless, when Claude Brosse reached the gates of the royal palace, he was disdainfully turned away.

A pious man, he stopped to pray at the chapel of the Château de Versailles—where the king himself was at prayer. At the moment of the elevation of the host, the king angrily noticed Brosse, whose head was high above those of the kneeling faithful. Shocked by such an apparent lack of piety, the king told one of his aides to order the man to kneel down—only then realizing that Brosse was indeed, kneeling, but was so tall that he looked as if he were standing up.

(Photos Bernard Jourdes)

After the mass the king had the man brought to him, and Brosse promptly took this opportunity to praise the wine he had carted so far. Entertained by the man and the incident, Louis XIV tasted the wine, liked it, and adopted it for his court. Today, three centuries later, the prestige of Mâcon once again is high with wine lovers.

THE COTE CHALONNAISE

Further north, in the Côte Chalonnaise, it may look as if the vineyards have disappeared—but they have not. They must be looked for to the west of the main highway. Montagny produces only white wines. The local growers maintain proudly (and justly) that their wine leaves the mouth fresh and the head light. Montagny is, in fact, a sort of compromise between Pouilly-Fuissé and the distinguished Montrachet produced still further north. The vineyards here no longer roll down steep slopes but share the land with pastures dotted with white cattle —which give the famed Charollais beef—and with thick forests. Nestled in this territory are villages that have been sleeping for seven, eight centuries and more. Three districts produce Montagny wine: Buxy, Saint-Vallérin and Jully-les-Buxy. Further north is Givry, which produces a rich, straightforward red wine that was praised highly by good King Henry IV, a connoisseur.

Mercurey, a small hamlet which has given its name to wines grown around it, can claim, thanks to these wines, to be the capital of the district of Chalon. The red wine has a deep, powerful garnet color. The gusto of the Pinot wine plant is revealed fully in this delightful wine, full-bodied and tasty, capable of aging long and well. Rully, dominating the wine valley of the Saône river, produces a tasty white wine, which lends itself well to be made sparkling. Local gourmets recommend it to accompany the *pochouse,* a sort of stew made with carps, eels, pike, and perch, fished in the river and prepared with white wine and herbs.

LANDSCAPE IN BEAUJOLAIS. *(Photo Louis Orizet)*

VINEYARDS run right up to the castle foundations. Château Aloxe-Corton, the fermenting house of the château of Corton-Grancey, Côte d'Or. *(Photo Lapie)*

THE COTE D'OR

Past Bissey, which produces a pleasant white *Bourgogne Aligoté,* the visitor enters a wine district which many hold to be the greatest in the world. It is the Côte d'Or (Golden Slope), a low meandering range of hills along the western edge of the Burgundian plain, and part of the French *département* of the same name. The Côte d'Or, with its subdivisions of the Côte de Nuits in the north and the Côte de Beaune in the south, possesses a soil that is poor, pebbly and granitic, and this may be why Benedictine monks at the turn of the millennium planted it with vines—it seemed unsuitable for any other culture. As it turned out, the composition of the soil, the drainage, and exposure to the sun, combined to yield magnificent wines.

The Côte De Beaune

Coming from Beaujolais and driving north, the visitor will first encounter the Côte de Beaune in Santenay, a spa with sulphured water said to be capable of relieving gout and arthritis. White wines (from Chardonnay and Pinot Blanc) and red wines are made here, in which a touch of the greatness of the Côte d'Or can already be detected.

Two miles due north, there lies a vineyard which produces, according to many wine lovers, the world's greatest white wines, derived from less than 20 acres of Chardonnay plants. It is the vineyard of Montrachet, partly within the limits of the township of Chassagne-Montrachet, partly in that of Puligny-Montrachet. Golden pale, with incomparable breed and supreme elegance, Montrachet has finesse and lightness, combined with body, and a bouquet with a touch of almond. "One should kneel down and uncover one's head to drink Montrachet wine," wrote Alexander Dumas. French wine lover and expert Maurice des Ombiaux describes it as having somehow absorbed more sunshine than any other wine: "Its smoothness lines the mouth with caresses, then its bouquet asserts itself with extraordinary strength and gentleness, like a *magnificat* under the vaults of a Gothic cathedral."

Two close neighbors are the vineyards of Chevalier-Montrachet and Bâtard-Montrachet, producing wines that can rival Montrachet itself.

The townships of Chassagne-Montrachet and Puligny-Montrachet have larger vineyards, producing excellent white wines, comparable to but not as superb as those of Montrachet and its two princes—Chevalier and Bâtard. The best wines from these townships are labeled with the names of exceptional vineyards, such as Chassagne-Montrachet Grandes-Ruchottes or La Romanée, or Puligny-Montrachet Les Perrières, Les Pucelles, Les Charmes. These wines reach pinnacles of harmony, of elegance, of breed, and it is difficult to choose among them. Everything about them is subtle, delicate, well balanced. They are particularly good with hot hors d'oeuvres—and they should never, never be overly chilled and served in an ice bucket, lest their superb eloquence be muted.

Although Chassagne-Montrachet is known particularly for its fine white wines, more than half of the township's production is not white but red. While not ranked among the finest red Burgundies, a few of the growths can be exceptional. The best are sold with township and vineyard names.

Directly north of Puligny-Montrachet, another village famed for great white wines is Meursault, whose name is said to derive from the Latin *muris saltus*—"the leap of the mouse", though no one seems to know just why. Most of the vineyards are planted with Chardonnay grapes, but there is also some Pinot Blanc.

It is not easy, after attempting to describe Montrachet, to do the same for Meursault, except perhaps to say that the latter has, to a lesser degree, the virtues of the former. Occasionally, however, Meursault can equal and surpass its neighbor, particularly when the wine issues from such vineyards as the Goutte-d'Or, Les Genevrières, or the tiny Sous-le-Dos-d'Ane (covering only seven acres).

These great wines are made according to centuries-old traditions. After the pressing, the wine goes to a temporary home, 60-gallon oak *pièces,* where it continues slow fermentation in a cool temperature of about 60 degrees. It will stay in this home until the spring; every week the cask is rolled about to mix deposits with the wine for, as the Meursault saying goes, "It is the lees (deposits) that

IN THE TASTING CUP,
the wine sparkles like a jewel.
(Photo Boubat, Réalités)

Opposite page:
COURTYARD OF
THE HOSPICE OF BEAUNE,
Côte d'Or. *(Photo René-Jacques)*

feed the wine." Only in spring are the lees separated from the wine, and the wine returned into clean casks to age for another two years before being bottled.

Further north, toward Beaune, the vineyards of Pommard (some of which date from Roman times) are among the largest in Burgundy, and its wines among the best known. Pommard wines can be light though full-bodied, and sometimes may require the addition of sugar to the fermenting must. This is the carefully controlled operation called chaptalization, practiced only in those years when there is not enough sunshine for grapes to reach their full maturity. Volnay nearby produces wines that may be lighter, but have a full bouquet.

BEAUNE, THE METROPOLIS OF WINE

The town of Beaune itself has given its name to several well-known surrounding growths. Center of viticultural research and former residence of the dukes of Burgundy, Beaune attracts many visitors with its churches and fortifications, its ancient streets and carriage gateways, its tranquil, old-fashioned medieval charm, its endless and hospitable cellars. One of its most famous landmarks is the Hospices de Beaune, hospital and shelter for the poor, financed in part by sales of wines. The Hospices, also called Hôtel Dieu, were built in 1443 by Nicolas Rollin, chancellor (and tax collector) of Burgundy, and his wife, Guigone de Salin. Rollin's generosity has permitted the poor to find free shelter and treatment here for more than 500 years. According to the 19th-century writer Stendhal, Louis XI commented pointedly that "it is only proper and fitting that Rollin, after having created so many poor, construct a hospital to house them."

Every year, usually on the third Sunday of November, there is an auction sale of wines for the Hospices, a *vente à la chandelle*—the burning of a candle setting the time limit for auctioneers. Extraordinary prices are sometimes paid.

Festivities in Beaune go on for three full days, and the motto which can still be deciphered on the facade of an old house in Savigny is scrupulously followed:

> *Il y a cinq raisons de boire:*
> *L'arrivée d'un hôte*
> *La soif présente*
> *La soif future*
> *La qualité du vin*
> *Et toutes celles qu'il de plaira d'imaginer.*

This sentiment parallels that of Henry Aldrich, a 17th-century English clergyman, who may have visited Burgundy and has consoled generations of wine drinkers with the following verse:

> *If all be true that I do think,*
> *There are five reasons we should drink:*
> *Good wine—a friend—or being dry—*
> *Or lest we should be by and by—*
> *Or any other reason why.*

The village of Aloxe-Corton marks a fitting end to the parade of vineyards of the Côte de Beaune. The red wines of the Corton vineyard, made from the Pinot grape, are the best of the Côte de Beaune. They are fine, delicate, yet robust and full of flavor. White wines (from the Chardonnay grapes) are comparable in quality to the Meursaults; the finest among them is Corton Charlemagne, grown in vineyards said to have been owned (and perhaps even planted) by the emperor himself. They have a golden color which feasts the eye, are rich in alcohol, and are distinguished by a scent of nutmeg.

Château Grancey, owned by winegrower and merchant Louis Latour, is one of the best-known domains of the Côte de Beaune. The winehouse itself stands at the foot of a hill covered by the celebrated Charlemagne vineyard, and the cellars are deep underground. A steep, stony spiral staircase leads to a sort of crypt, used as a wine-tasting room. The walls are covered with *Racodium cellare,* the tiny mushroom which makes a colorful, abstract design, and also helps regulate the humidity by absorbing moisture. One of the best red wines to be tasted here is the Château Corton-Grancey, produced from vineyards directly overhead. The best wines of the district are labeled Corton, sometimes followed by the name of a vineyard. Less distinguished growths, as well as some wines from the adjoining villages of Ladoix-Serrigny and Pernand-Vergelesses, are labeled Aloxe-Corton.

The Côte De Nuits

Beyond Pernand-Vergelesses and Ladois-Serrigny, and past the stone quarries of Comblanchien, begins the realm of the great red wines of the Côte de Nuits, named for the town of Nuits-Saint-Georges. The map of the area reads like a rare wine list on which, within less than 10 miles, lie some of the world's most famous vineyards. It is here that the notion of *terroir,* the parcel of land covered by a particular vineyard, takes on its fullest significance. Sometimes the eye can see no difference between one parcel and the other, and only a narrow path indicates the borderline between one vineyard and the next. Sometimes, there is no visible separation at all—and yet the palate can discover, between the wine

LINED UP LIKE SOLDIERS, these glasses will toast the great event. *(Photo Brassaï)*

of one *terroir* and that of another, a wealth of nuances difficult to describe—and even more difficult to account for.

Nuits-Saint-Georges is a small town in whose cellars great wealth is hidden, for it is here that merchants and growers accumulate, and age, most of the wines produced on the Côte de Nuits. In 1934, the *Confrérie des Tastevins*, the world's best-known fraternity of wine lovers, was born here. It has branches in the United States and England.

The most subtle, delicate, feminine red wines of the Côte de Nuits come from the vineyards of Vosne-Romanée, sometimes referred to as the central pearl of the Burgundian necklace. The Romanée and Romanée-Conti vineyards are the most distinguished ones here—yet nothing visible separates them from their neighbors, and the mystery of their extraordinary wines remains unsolved. Madame de Pompadour favored Romanée wines above all others, and tried, by wile and guile (but unsuccessfully), to dispossess the owner of the vineyard, Prince de Conti, who kept most of the wine for his own use and for his friends.

Romanée-Conti wines have a bouquet with a spiciness that has been termed oriental, with a touch of violet and cherry, a clean, deep ruby color, and a caressing texture. Great and costly efforts were made until World War II to preserve the pre-phylloxera vines, ungrafted on American root stock. Replanted in 1946 with grafted vines, Romanée-Conti now produces wines said to be close in quality to earlier vintages.

Comparison of Romanée-Conti with its distinguished neighbors is difficult, and most wine lovers agree that it is better to give to each its own due. Fortunate are those who can have this chance: The annual yield is some 500 gallons at Romanée, 1,250 gallons at Romanée-Conti, 2,700 at Romanée Saint-Vivant, 4,500 at Richebourg (a favorite of Ernest Hemingway), 3,200 at La Tache, 3,700 at Grands-Echézeaux, in Flagey-Echézeaux nearby, and perhaps as much as 15,000 gallons at Echézeaux itself—not much for a world of wine lovers. Vineyards entitled to the Vosne-Romanée place name produce altogether 100,000 gallons of wine a year—and though most do not bear the names of any of the district's most famous vineyards, they are all uncommon wines.

Next to the vineyards of Grands-Echézeaux is the Clos-Vougeot, Burgundy's largest vineyard (some 125 acres). It was planted in 1111 by Cistercian monks who also built a wall around it, making a boundary that stands to this day. The monks soon noticed that there was a difference in the wines that issued from different parts of the vineyard—notably between those of the upper and those of the lower slopes. Following the hierarchy to which they were accustomed, they produced three different growths: the *cuvée* of the pope, the best, from the upper part of the enclosure; that of the king, almost as good but from the middle part; and that of the monks from the lowest part. The monks kept control of the vineyard until the Revolution, building up its reputation as one of the region's finest. So great was its renown that one of Bonaparte's officers, a Colonel Bisson, while marching past it to join his chief in Marengo, stopped his regiment and had his men present arms before continuing on his way.

CELLARS OF THE CHATEAU DU CLOS-VOUGEOT, Côte d'Or. *(Photo Bernard Jourdes)*

NAPOLEON TOASTS HIS GUESTS. Chambertin was his favorite wine. Imagerie de Paris, *Bibliothèque nationale,* Paris. *(Photo Ph. Brossé)*

Napoleon himself, though no great wine lover, appreciated Clos-Vougeot as well as Chambertin, but is said to have drunk his wine mixed with a little water. He is also said to have received a message from the last clerical cellar master, appropriately named Dom Goblet, to the effect that the cellars held some forty-year-old Clos-Vougeot. "And if he wants to drink some," added the crusty cellar master, "let him take the trouble to come here." Napoleon didn't come.

Dom Goblet's portrait presides over the imposing Renaissance fireplace of the massive Château du Clos-Vougeot. This building, begun in 1551 by Don Jean Loisier, then abbot, is a solid, proud and stocky structure, now owned by the *Chevaliers du Tastevin.* Several times a year these wine lovers gather there to mix good food, good wines, warm fellowship and hospitality, and rousing drinking songs. Restored extensively after World War II it is an imposing edifice that still contains wine presses hewn from great oak trees around the year 1000.

Owned nowadays by several dozen people, the Clos-Vougeat vineyard is surrounded by a ten-foot-high wall of stone. The wine itself can vary greatly, from parcel to parcel and grower to grower—but can be safely ranked among the great wines of Burgundy, having a bouquet reminiscent of truffles and violets.

Still further north, a mere few hundred yards away, Chambolle-Musigny produces wines of great distinction. A typical sample was described by Jules Chauvet, an expert in the matter, as "having a magnificent color, ruby rather than garnet, a basic aroma of lichen, of sandalwood, of cherry, resin, venison musk, gray amber, Russian leather, jacinth, narcissus, strawberry and Havana tobacco. The taste is supple, powerful and long-lasting, the harmony is total, and elegance supreme."

The two best vineyards in the Chambolle-Musigny township are Musigny and Bonnes-Mares. The Musigny wine, from a vineyard right next to Clos-Vougeot, has breed hardly ever surpassed by any wine, and equaled by only a few. Bonnes-Mares, a 37-acre vineyard, mostly in Chambolle-Musigny, partly also in Morey-Saint-Denis, is said to have more power but somewhat less finesse than Musigny—ranking nevertheless in the top dozen of the most superb Burgundy wines. Among other renowned vineyards in the same district are Les Amoureuses, Les Charmes, and Derrière-la-Grange ("behind the barn").

Morey-Saint-Denis (the village of Morey having added the name of one of its famed vineyards, Clos-Saint-Denis, only in 1927) has vineyards dating back to the Roman period. The best wines of this township come from two vineyards, Clos-de-Tart, and Clos-des-Lambrays. Clos-de-Tart was the property, eight centuries ago, of Bernardine nuns, and today is owned by a single proprietor, Jean Mommessin, grower and merchant—a rare occurrence in the Côte d'Or, where most vineyards are shared by several owners. As a result, all bottles of Clos-de-Tart are certain to be the same wine, made and bottled by the same winery. It is one of the most full-bodied, long-lived wines of the Côte d'Or. Clos-des-Lambrays wine has long been famed for its strength and full body (in 1385, Marguerite of Burgundy drank it to accompany pheasant sparkled with gold dust, pullet with saffron, and meatballs made with venison and red currants).

At this point, a visitor to the Côte d'Or should uncover his head (as he already has for Montrachet) to meet one of the world's most superb red wines, Chambertin. It comes from vineyards first laid out by a farmer, Bertin, who in the 13th century wanted to make wine as good as that produced by the Cistercians monks nearby. The vineyard, first known as Bertin's field *(champ de Bertin)* became so famous that its predecessor and rival, the Clos-de-Bèze, planted in A.D. 630 by the monks of the Abbaye de Bèze, added Chambertin to its name.

With grace and vigor, finesse and delicacy, Chambertin and Chambertin Clos-de-Bèze vineyards (covering, respectively, 32 and 37 acres) synthesize the virtues of the wines of Burgundy.

French schoolchildren know that Chambertin wine was Napoleon's favorite, and later learn that Alexandre Dumas wrote: "Nothing makes the future appear so rosy as to contemplate it through a glass of Chambertin." Talleyrand also praised the wine, and was adamant that it be drunk with proper respect. "Monsieur," he once told a guest who was drinking his Chambertin without enough consideration, "when one has the honor to be served such a wine, one takes one's glass with much respect, one looks at it, one takes a whiff of it and then, having put it down on the table, one talks about it!"

Escorting Chambertin and Clos-de-Bèze, a court of worthy princes follows: Charmes-Chambertin, Ruchottes-Chambertin, Mazis-Chambertin, Chapelle-Chambertin, Griotte-Chambertin, Latricières-Chambertin, Mazoyères-Chambertin, Cazetiers, Clos-Saint-Jacques, Etournelles, Fouchères, and Varoilles. Lesser nobility join the parade, under the district name of Gevrey-Chambertin.

A few more vineyards, a few more names—Fixin, Marsannay-la-Côte, Chenove,

Conchy—and the Burgundian miracle comes to an end at the gates of Dijon, capital of the Côte d'Or département and France's most renowned gastronomical center. Here stops the kingdom of Pinot, where this noble plant has surpassed itself. It is perhaps to mark the limit of this land that a Burgundian called Noisot, commander in Napoleon's Imperial Guard, ordered a bust of his emperor to be erected in his park. In the twilight of his career, having returned home from the wars, Noisot felt the time had come for him to meet his emperor once more—but in another world. He ordered that he be buried standing up, saber in hand, facing Napoleon's statue. Such a wish might appear extravagant—but many similar testimonies to the grandeur of the Burgundian heart can be found, carved on tombstones in this magnificent country.

Witness the following epitaph, engraved in stone, somewhere between Chambolle and Gevrey:

> *Here rests*
> *François Gabriel Charavin*
> *Retired captain*
> *Member of the Legion of Honor*
> *Born in Gevrey, February 9, 1785.*
> *He served his beautiful country faithfully*
> *From 1809 to 1846.*
> *He made war in seven kingdoms*
> *Took part in twenty battles,*
> *Eleven skirmishes, one blockade, three sieges*
> *Under the reigns of Napoleon, Louis XVIII, Charles X*
> *And Louis Philippe.*
> *He died on December 16, 1870*
> *At the Battle of Nuits*
> *Defending his vineyards.*

HANDLE OF A WINE-TASTING CUP.
Musée des Arts décoratifs, Paris. (Photo Chapman)

CHABLIS

Some sixty miles from the Côte d'Or, toward Paris, lies the wine-growing region of Chablis—whose history links it to Burgundy, though its geography would put it in the Ile de France, with Paris, or even in the Champagne region. That Chablis opted to be part of Burgundy, however, makes good sense oenologically and gastronomically.

Before the railroads criss-crossed France, the vineyards of Chablis covered more than 100,000 acres. Now, there are less than 20,000 acres left as a result of competition with cheaper wines that have been reaching Paris. It is a pity to behold once prosperous vineyards turned into pastures.

According to the rules of *appellations contrôlées,* Chablis Grand Cru and Grand Chablis must be made exclusively from Pinot Chardonnay grapes, and from the best vineyards in the townships of Chablis, Milly, and Poinchy. Chablis Grand Cru comes from a single slope, facing the town of Chablis across the Serein river, and subdivided into seven small vineyards: Les Clos, Blanchot, Valmur, Grenouilles, Vaudésir, Preuses, and Bougros. Twenty other small townships are allowed to label their wine as Chablis, the total annual production being around 350,000 gallons. In addition, some 100,000 gallons of wine entitled to the name of Petit Chablis are produced in lesser vineyards, also from Pinot Chardonnay. Most of these are drunk as plain table wine, usually served in a decanter, when less than a year old.

Chablis wines are among the most pleasant white wines of France, unpretentious and fresh, neither sweet nor too acid, light but not weak (they contain at least 10 percent alcohol) nor dull. Clear and translucent, they are said to be distinguished by their hygienic and digestive qualities and by the lucid exaltation they give to one who drinks them.

Now taking leave from Burgundy, the visitor should do it *tastevin* in hand, in the traditional Burgundian fashion. This artfully embossed cup, preferably made of silver, is transmitted from father to son, and has been used to raise countless toasts to the success of the greatest vintages of Burgundy. Renewing this gesture when parting from their guest, Burgundians fill their *tastevin* with wine, but also with their happiness and friendship—hoping the visitor will take some of these blessings along.

PORTABLE WINE BARREL for the use of vintners during the harvest. *(Photo Chapman)*

THE MARNE flows through the gentle slopes of Champagne. *(Photo Bernard Jourdes)*

CHAMPAGNE

The mere word "Champagne" suggests an iridescent bubble filled with luxury, with pleasure, and with sparkling elegance, reflecting the most charming attributes of an entire civilization. The writer's pen itches for glorious epithets—but let it pause, for the word Champagne covers a vast number of wines, ranging from poor to excellent, that have nothing to do with that section of France that answers to the name Champagne. Even the Soviet Union, that most avowedly proletarian of republics, produces vast quantities of wine that imitates this symbol of aristocracy.

THE LEGEND OF DOM PERIGNON

According to legend, the development of Champagne came about thanks to a 17th-century monk, Dom Pérignon. It was through his inventiveness and his refinement of taste that the men of Champagne learned to blend their wines and give them a delicacy that they had never had before. Add to this the know-how and perseverance of the great landowners, who wanted their wines to compete with, and outstrip, those of Beaune, and the wines of Champagne were on their way to becoming the wines we know today.

Of Dom Pérignon, Champagne's inventor, there remains but one memoire, written not by the dom himself but by his pupil and successor, Brother Pierre of the Abbey of Hautvilliers. The title of his treatise, *Of the Culture of Champagne Vines Located at Hautvilliers, Cumières, Ay, Epernay, Pierry, and Vinay,* gives an idea of the extent of the abbey's holding, and thus of its wealth and influence. The chronicle tells of the founding of the abbey:

"Toward the middle of the 7th century, if one believes tradition, St. Nivard, Bishop of Reims, and his nephew, Berchier, planned to found a monastery on the banks of the river Marne. The walk was long, the day was hot, and the two holy men were tired. They sat down and St. Nivard promptly went to sleep. He saw then, in a dream, a dove flying away and sitting on a tree—and St. Berchier saw the same dove, though he remained awake." This was a good omen, and the two saints selected this place as the site for their new abbey, which was to become famous as the Abbey of Hautvilliers.

That fame it owes to Dom Pérignon, whose religious superior wrote: "This precious man is forever dear to the land of Champagne for having made wines reach the degree of delicacy, of reputation and of vogue, where we see them to be now, and dear to France herself for having so spread this first branch of her trade, and so increased her wealth." Other writers even claim that the use of cork in closing the bottle, wine making in cool cellars, the shape of the bottles, the tying of the cork, and the long tulip-shaped crystal glass—the flute—come from Dom Pérignon's initiative.

Before Dom Pérignon, Parisian merchants were buying the "little greenish wines" of the Champagne area only because they were cheaper than Burgundy. Then the landowners learned to treat their vineyards with such judicious care that the wines of the region became "light and a bit feline, subtle and delicate and of a pleasant taste to the palate, and hence wished for by kings, princes, and lords." With the help of some favorable propaganda by the physicians of King Henry IV, and the opportunity of serving Champagne at one feast or another, it took less than a century for wines from such important Champagne centers as Ay and Sillery to bring prices four times that of the wines of Beaune.

"I have never seen anywhere anything that could be compared to the care and precautions taken by the *Champenois* in the past fifty years," wrote Abbot Pluche about 1720. "Their wine was, before that, very fine and esteemed, but it did not hold well and did not travel far. Through long experience, they have succeeded in making it firmer to the point that it holds much longer, without losing anything of its pleasantness."

The practice of making white wine from both white and black grapes was then well known, and the wine they made was probably *crémant*—slightly more sparkling than natural still wine, but far from the subtle sparkling wines we have today. Before the turn of the 17th century, the French already knew and liked a "devil's wine," which is believed to mark the appearance of truly effervescent champagne. Then, under the tutelage of Dom Pérignon and for 47 years until his death, the growers of Champagne learned "to choose the vine plants suitable to the soil, to layer them into the earth, to prune them, and to mix the grapes."

THE CHAMPAGNE VINEYARD

The wines the Abbot Pluche described came then, as they come now, from a region so close to Paris that they have been called simply the wines of France— that is, the Ile de France. The vines begin at Chateau-Thierry and follow the valley of the Marne, along charming hillsides reflected in the river, to the towns

A POSTER BY BONNARD, "France-Champagne," 1889.
Bibliothèque nationale, Paris. (Photo Giraudon)

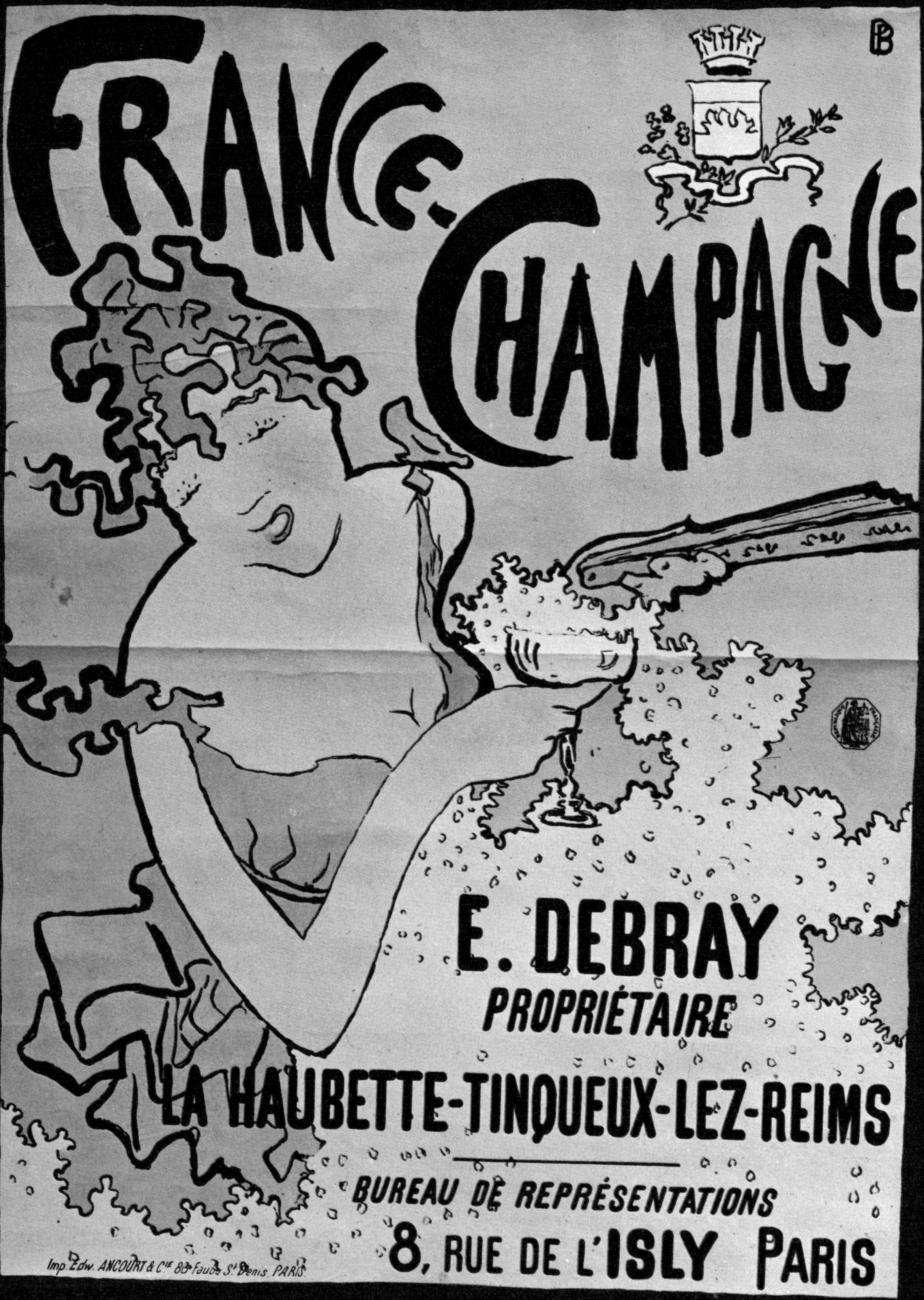
FRANCE-CHAMPAGNE
PB
E. DEBRAY
PROPRIÉTAIRE
LA HAUBETTE-TINQUEUX-LEZ-REIMS
BUREAU DE REPRÉSENTATIONS
8, RUE DE L'ISLY PARIS
Imp. Edw. ANCOURT & Cie 83 Faubg St Denis. PARIS

of Epernay, Ay, Avenay, Bouzy, and Ambonnay. South of Epernay lie the hills known as the Côte des Blancs, the "Slope of the Whites," planted almost exclusively with vines of the Chardonnay variety, which bear white grapes. The important towns are Cramant, Avize, Le Mesnil-sur-Oger, Vertus, and Bergère-les-Vertus. (Bergère means "shepherdess," and when foot soldiers of World War I marched through this land, they sang, "Bergères, there's hardly any, and virtues, there are no more.") North of Epernay, the *Montagne de Reims,* crowned by forests, is planted with Pinot Noir, and is called locally, the *Montagne de noir,* the "mountain of the black."

Then come Mailly, Verzenay, Verzy, and Villers-Marmery; the countryside unfolds in gentle vales with green slopes studded with church steeples. There are modest hillocks with oak forests protecting old ruins, towers, and hollow paths. This is the country of jugglers, fables and fireside stories. It is the northernmost vineyard of France, and the vine is demanding, requiring precise and thorough pruning, regular plowing, dressing, weeding, paring, stripping of leaves, and protection with earth.

All along this thin strip, nearly 80 miles long and seldom wider than a mile, some 30,000 acres of porous chalk lends itself to no other culture than that of the vine. This is the Champagne vineyard—less than one hundredth of France's vine-planted land.

The growers of Champagne, while keeping pace with new techniques, still follow the precepts of the legendary abbot, to bring to the presses "with gentleness and cleanliness" bunches as intact as possible, so that they retain within them all the fragrance of fresh grapes.

WITHOUT CORKS, CHAMPAGNE COULDN'T EXIST.
(Photo Jahan, Plaisir de France)

The harvest is carried in great "mannikin baskets" each holding up to 200 pounds of grapes. These are loaded onto chariots (still often pulled by horses) and driven to the nearest *vendangeoir,* a sort of a preliminary winery where grapes are piled as if for a Gargantuan dessert. The harvesters often select the handsomest bunches to go on top, for no other reason than to please their eye and tickle their pride. The sale of the harvest usually takes place here—traditionally it only becomes binding the moment the buyer has his grapes poured into the *vendangeoir's* press. Pressing is done by a *pressureur,* a specialist usually independent from the grower as well as from the winehouse, whose job is strictly legislated, and who keeps records of the weight of grapes, sugar content, and the amount of must obtained. Pressing is usually done in batches of 8,000 pounds of grapes, each being termed a "marc."

The harvest must never be crushed: the rubbing of the skins together would invariably give color to the juice. The harvests from different parcels are never mixed: the growths remain separated until the later stages, when the blend is prepared. It is considered to be better, before any mixing takes place, to make sure that each growth develops into a clear wine, capable of standing up by itself before a critical and thorough wine tasting. This is done to limit any unforeseen event that could endanger the first fermentation of the wine. Each 8,000-pound marc is pressed three times to yield about ten *pièces* of about 54 gallons each, which are reserved for the *cuvée*—the must that will become Champagne. Between each pressing, the *pressureur* and his assistants open the press and cut the marc with wooden shovels. Then one or two more pressings will give another 80 gallons of must which can be added to the *cuvée* of certain years. Another pressing can yield 80 more gallons, but by now the must is poor in sugar and in alcohol, too acid and too full of tannin; this part of the pressing is not entitled to become Champagne, but will be mixed with the small table wines of the Marne.

In the old days, the big wheel of the press required the work of no less than seven men. Today, the Wilmes press and electric power have simplified this operation. The must, with a slight addition of sulphur, is poured into 4,000-gallon vats, in which it rests for a dozen or so hours to let the most important part of its impurities settle at the bottom, a method considered to be better than clearing by centrifuging or filtration. Sometimes the must is slightly sugared, so that the wine can reach a minimum of 10 to 11 percent alcohol.

FROM FIRST FERMENTATION TO BLENDING

The must now goes to the central buildings of the large Champagne companies, where the first and tumultuous fermentation goes on in glass-lined metal vats or vats of oak. For two or three weeks, the cellars dug in chalk must be kept at a temperature of 60 to 65 degrees, for overheating might disturb the fermentation. Sometimes the metal vats must be sprayed with cool water, and the large companies have now installed—or plan to—complete air-conditioning systems in their wineries. The cellars under the towns of Epernay and Reims, not unexpectedly, are quite vast. The chalky ground is wormed with galleries whose length may well be equal to that of main roads in the region, and certainly exceeds one hundred miles. (The Moët et Chandon cellars at Epernay alone consist of some 15 miles of galleries.)

At the beginning of December, the yeast cells have finished their work and settle down to rest at the bottom of vats. The wine is racked (a process akin to decanting) several times, until all solid matter that would be damaging to it is separated.

The Champagne-to-be is now in casks. At the foot of each cask, owner and cellar master meet for conferences, to give each wine an exhaustive tasting, and to select the first assemblage of different casks of the same growth to be mixed in a vat.

The most solemn moment of the year comes at the end of January. After another racking, done with the help of a specially designed funnel with a valve, the wine makers must decide which growths should be blended: the wines must be "established." The expert's nose and palate go to work again, to accomplish a task that dates back to the times of Dom Pérignon. To him it is said "a basket of grapes was presented that had been picked in all the vineyards of the abbey and of Cumières; he tasted them, and put them all aside, according to the soil from which they came, and marked with much assurance the grapes that should be blended to have the wine of the best quality." Perhaps today the task is easier, for the blending is made with finished wines, and not with unpressed grapes. When it comes to this point, each Champagne company has its own secrets and tries, from year to year, to produce the best type of mixture while keeping constant the character which has become known to its customers.

Since nearly every Champagne is a blend, the brand name, in this case, is more important than the vineyards or districts from which the wines have come.

Sometimes wines from another year are added, for they can give support and improve the current growth. Traditionally, the proportion of older wines added did not exceed a third of the total volume, since a larger proportion of aged wines threatened the development of the sparkle during the second fermentation. Today, the fermentation process being better understood, any portion of old wine can theoretically be added. Only the vintage years remain unmixed with older wines, so as to keep the character that makes the year particularly good. The vintage years in fact are frequent, recently occurring not quite every other year: 1945, '47, '49—all three now almost impossible to find—1952, '53, '55, '59, and 1961 and '62.

Champagne is made with both white and black grapes, but there are more of the latter, proportions wavering around three to one. Black grapes are the Pinot Noir or Pinot Meunier, coming from the Ay district for breed, finesse and fullness; from the slopes of Bouzy and Ambonnay for body and support; from the *Grande* or *Petite Montagne de Reims* for fragrance and for freshness.

One fourth of the harvest is the white Pinot Chardonnay, which adds some freshness, brings elegance and grace and delicacy, and foams without requiring any prompting. Little can match the sheer delight of a pure *blanc de blanc* made with Chardonnay grapes only. If the wine is of a good year, it covers the tongue and the palate with fresh lacework, delicate and perfumed, that makes one want to dance. But such wines are rare. The public taste prefers that at least one fourth of a wine be made from the black grape. The *blanc de noir* wines, made only from the dark grape of Pinot, are full and rich in alcohol, but lack the charm and delicacy that emanate from the white grape.

In vintage years, no more than four fifths of the yield can be sold under vintage label. Vintage Champagne is, in addition, subjected to an expert wine-tasting committee that must give its approval before the wine is shipped, at least three years old, as vintage wine.

Since 1920, the best growths are rated to a maximum of 100, and they will not pass the test if grades are below 70. (A similar rating is used to determine the price of the grapes.) Not all excellent Champagne, however, is sold under the brand names of the well-known companies, such as Moët et Chandon, Veuve-Cliquot, Bollinger, Roederer, Taittinger, Lanson, Pol Roger, Perrier-Jouet, Mercier, Pommery or Mumm. A few interesting and sometimes very good Champagnes are sold under the name of their district of origin; these are made by small producers, and are usually unblended. Thus excellent wines go under labels of Cramant, Avize, Ay and Sillery, among others.

Finally the *cuvée* is born, surrounded with much anguish and many precautions. At the end of the winter it is given another analysis, as well as some tannin to immunize it against a disease called *graisse,* caused by bacteria, which give the wine a greasy or oily appearance and consistency. It receives pure, selected yeast cells to start the second fermentation, and small amounts of sugar.

SECOND FERMENTATION—ADDING THE SPARKLE

A good sparkle requires that the wine contains six times its volume of carbon dioxide. This will be made possible by added yeast and sugar, which will

A HARVESTING BASKET, typical of the Champagne region. *(Photo Bernard Jourdes)*

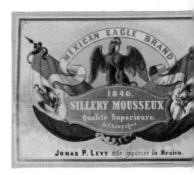

bring the wine to a pressure of six times that of normal atmosphere at sea level. The foam of good Champagne must be light, lively, and long-lived, and the bubbles must be small and rise continuously and tirelessly from the bottom of the glass. Only excellent grapes, subjected to a treatment beyond reproach, will give this result, whose underlying causes are not completely understood. When the foam is rough and rapidly falls flat, the exquisite refined sensation given by a true Champagne is lost: good Champagne "holds its foam"—and connoisseurs will drink it sip by sip, never in a long draught.

Now spring has come and the wine must be bottled. The wine is pumped out of its cask along clean, shiny pipes to the bottle filler. Filled and corked bottles, with the thick cork tightly wired to the bottle's neck, are then stacked away for the second fermentation to take place unhurriedly over several months. During this period the wine develops its carbon dioxide, as well as small additional amounts of alcohol. The bottles must be expertly stacked, thousand upon thousand, and neatly separated by thin laths—so that if one of them explodes, others around it will not be disturbed.

Three or four months later, the wine has built up its desired gas pressure. If seen against a candle, the bottle shows that a deposit has accumulated all along its lower flank. The bottles are then restacked, and as each one is moved from one pile to the other, it is expertly shaken after being marked with chalk, so that it is replaced exactly in the same position, and the deposits form again on the same side—but homogeneously this time.

THE BOTTLE "DANCE"

The wine rests thus two or three years—then it moves to yet another part of the cellars, into orderly ranks of specially designed racks—invented, it is said, by the *veuve* (widow) Cliquot, ancestral owner of one of the largest Champagne companies, Veuve-Cliquot. These racks are oak "pulpits," to hold the bottles at an angle of about 30 degrees to the vertical, neck down. And here begins the Champagne bottles' "dance"—the *remuage*. The bottles are standing on their heads and *le remueur* comes every day to give them a slight twist—a quarter turn

19TH-CENTURY CHAMPAGNE LABELS, except the one at far left, which was a fraud even at that early date. *(Photo Chapman)*

or less. At the same time, he slightly straightens up the bottle, bringing it closer to the vertical position. This goes on, three times a week for three months. Gradually, all sediment remaining along the bottle's wall is shaken off and it slides down, accumulating against the cork. A good *remueur* can thus manipulate up to 30,000 bottles a day.

Finally the wine is "candle clear"—except for a thin layer of sediment that adheres to the cork. The trick is to remove it by *dégorgement*—disgorging.

The bottles are gently wheeled into a workroom, where rows of machines wait. The neck of the bottle is frozen and the deposit becomes hard. The wire holding each cork is snapped, the bottle sneezes, and the cork shoots out—with the deposit—to be replaced by a temporary cap. This used to be done by hand—a single *dégorgeur* could handle a thousand bottles a day, losing, every time, no more than a teaspoonful of the precious liquid.

Now the bottles travel to yet another machine for "dosing." Dosing consists of the injection of a small amount of cane sugar, dissolved in Champagne brandy. The amount of this liqueur to be introduced in the wine depends on the type of Champagne that is sought: very dry *(brut)*, extra dry, dry, semi-dry, semi-sweet, or sweet.

Theoretically, Champagne Brut contains no added liqueur at all, and is completely dry, or natural. This is no longer so; usually a small amount of liqueur, up to one half percent of the bottle's content, is added. One to two percent go into extra dry, three to five percent for dry—also known as *goût Américain*, six to seven percent for semi-dry, and eight to twelve percent for sweet.

If too much empty space remains in the bottle, it is refilled with wine from the same growth. The bottles are then corked. The cork is made of several crisscrossing layers to prevent gas—compressed to a pressure of up to 75 pounds per square inch—from seeping through, and looks, when new, as if it were four times too big to enter into the bottle's neck. Thus one may wonder, examining a cork taken from a bottle of Champagne, why it looks like a sort of mushroom with a bell-shaped stalk. This is because a supple cork was chosen for a wine to be drunk in the next year or two. A cork much harder is selected for a wine to age

—and the harder the cork, the thinner the mushroom. The upper part, protruding from the bottle, expands into a sort of bulge, not unlike a rotund stomach escaping from its corset. And this is why, perhaps, the custom is to "dress" this bulge. The bottle top, thick and heavy to resist pressure, is given a sort of helmet, held by wire, to hide these fatty shapes. Before the bottles are so dressed, they are once more put down to rest for a few months, and then another screening eliminates all that may still have a shade of deposit.

THE WORLD'S MOST VERSATILE WINE

Before a bottle of Champagne goes on the market, how much care, work, and long unproductive storage are involved! This is why Champagne is not cheap, even in its country of origin. It is particularly not cheap in the United States, where sparkling wines are subjected to an exorbitant tax—twenty times that for still ones. On the other hand, if a bottle of Champagne is defective or mediocre, the purchaser should never hesitate to claim the return of the price he has paid for perfection.

The reputation of a house is usually the best guarantee of good Champagne. However, a grower who makes his own wine can sometimes offer at half the price of a better known brand a truly excellent Champagne, well made, and made with love. Look in a corner of the label for the two significant letters, R.M., for *recoltant manipulant*, meaning roughly "made by the grower." And after a few trials you may discover the rare bird. The choice is wide: 8 growers out of 10 work less than three acres of vineyards, and 2,700 families have properties between three and a dozen acres. Vineyards exceeding a dozen acres number no more than a hundred.

Until about 1750, the sparkle in wine was not considered worthy, and connoisseurs preferred still wine. But then the French and English, beginning with their kings and princes, took a great liking to sparkling Champagne. Perhaps on the advice of Madame de Pompadour, who liked Champagne "because it leaves you beautiful after you drink it," Louis XV in 1728 authorized the town of Reims to ship its wine in bottles—a practice that had been forbidden to limit fraud. A nephew of Dom Pérignon founded, in Reims, the oldest Champagne house. His example was promptly followed by others. Claude Moët, wine merchant since 1716, settled in Epernay in 1743 and began a tradition of excellence that has lasted to this day—thanks largely to the efforts of the *Comité interprofessionnel des vins de Champagne*, a self-policing body created at the initiative of the Moëts.

Every large Champagne house is proud to open its order book, upon which can be found, written in beautiful, large, cursive handwriting, the names of kings, princes, marshals and bishops.

Champagne is the world's most accommodating wine, for there is no menu, no gathering, no occasion and no hour of the day or night to which it cannot add its silky touch.

A GLASS OF CHAMPAGNE, showing the froth of fine wine. *(Photo André Causse)*

THE LOIRE VALLEY

To the Frenchman, the Loire Valley is something special.

It is, of course, the château country beyond compare, where French kings built elaborate and handsome castles along the tranquil river banks. It is a favorite fishing and hunting country. But, more than that, the Loire to some extent is France itself—unlike Brittany, Burgundy, or Provence, which are, first of all, provinces, and only afterward a part of France.

The Loire, where, in cities such as Tours, the French language is best spoken, is the essential France, with its moderate and reasonable clime, whether rainy or sunny, cool or warm, but never scorching or icy. It is France with its landscape of gentle slopes, of thick woods that abound with game, of rich pastures and fields and gardens, of secret rivulets and quiet rivers filled with fish, of the majestic Loire itself, sometimes capricious (as a French river should be). It is France with shady groves and quiet towns—a countryside dotted with castles of all sorts and sizes, huge medieval cathedrals and modest churches, as old as French civilization.

It is France and its vineyards, shared among many growers or divided into individual parcels—all forming part of the landscape, yet without overwhelming it. They are vineyards that bespeak patience, a quiet disposition, and the love of work well done. The people here have forged the French language, have written French history, and have, to a large measure, created the French way of life. This part of the French country, still the most faithful to religion and tradition, produces an amazingly varied garland of wines. Some are poor—others superb, and even great.

Near the headwaters of the Loire is the city of Vichy—the wartime capital of occupied France, famous mostly for its spa where tired city dwellers come to take the waters. Madame de Sevigné, noted letter writer of the 17th century, once commented: "One goes to the fountain at six o'clock; everybody meets there; one drinks and makes a face because—can you imagine!—the water is boiling and with a very disagreeable taste of saltpeter." The area also has its wines, however, some more of historical than contemporary interest, but at least one good enough to have been placed under French government controls.

The lesser wines are Chanturgues and Corent, the former so hard to come by that some have wondered whether it exists at all. Traditionally these wines were made from the Gamay grape, imported from Burgundy. However, since there are no legal controls over their production, it is theoretically possible to make them from anything. Of more interest is a curious wine known as Saint Pourçain. The quality of this is protected not by the laws of *appellations contrôlées*, but by regulations covering lesser wines, *vins délimités de qualité supérieure* (V.D.Q.S.)—regional wines of superior quality. Red, white and rosé wines are made, the reds and rosés from the Gamay grape, the whites from the little-known Tressalier which produces a light, dry wine with a taste reminiscent of tart apples.

Downriver from the mouth of the Allier, a tributary of the Loire, begins the area where the Sauvignon grape reigns supreme. Here, however, it is called Blanc

Fumé, which gives an indication of the color and taste of the wine it gives (*fumé* meaning "smoked"). The uppermost vineyards are those around the town of Pouilly-sur-Loire, producer of two wines, one pleasant enough, the other having some distinction. Pouilly Fumé (known as Blanc Fumé de Pouilly) is the good one—a lively dry white wine with a spicy scent and considerable finesse, and capable of extreme longevity. It should not be confused with the lesser wine, known simply as Pouilly-sur-Loire.

The greatest threat to vineyards in this area is frost, a humid frost that is apt to set in during the spring when the wind blows from the west. The trouble is compounded when the sun comes out, for it can burn the frost-bitten buds, leaving them little chance of survival. The winegrowers' defense is a set of burners—simple tin cans filled with wet straw and gasoline which, when ignited, give off a heavy smoke that screens the frost away. It is not uncommon for growers to leave their beds in the gloom of night to light their burners when frost threatens. The work is endless and arduous, but the rewards are great, for Pouilly Fumé is one of the more engaging of the white wines of France.

SANCERRE

Across the river is the area of Sancerre, rising from the old port of Saint-Thibaut, established by the Romans and still an important shipping center. Nearby stands a mountain, whose trees supply the wood for wine casks. The

HOMAGE TO VOUVRAY WINE, a turn-of-the-century postcard. *Massal collection.*

THE SQUARE TOWER OF ROCHECORBON.
Vines grow to the very foot of this tower overlooking the Loire. *(Photo Arsicaud)*

steep slopes of Sancerre face the sun, and the Sauvignon vineyards are longtime residents of the area. The ancient Abbey of Saint-Satur once owned large holdings here, and, according to one 12th-century source, "nearly all were planted with vines, from which a large amount of very good and excellent wines are made every year, and of which there is great commerce, by water and by land."

Since the French Revolution, the vineyards have been divided among many growers, most of whom own several often widely scattered parcels. The vines are crowded—more so than in most vineyards—but the harvest is not a large one, ranging from 150,000 to 200,000 gallons annually. Much of this is sold to cafes and restaurants in Paris, or directly to private customers.

A newly planted Sauvignon vine must wait eight years before the wine it yields becomes entitled to the name Sancerre (whereas most other vineyards graduate at their "fourth leaf"). And the harvest often does not begin before nights have grown cold, and there is no more hope for grapes to continue maturing. Once vats are filled, it may be necessary to heat the cellars so that fermentation can be completed by the beginning December.

Sancerre is a lively, fruity wine, very full-bodied, and of a beautiful pale yel-

low color that can turn to gold. It is almost invariably dry, sometimes a bit acid yet smooth, particularly in warm and humid years when some of the grapes have become affected with the "noble mold," as in Sauternes. The wine is made in 14 parishes, or towns. The best come from the Sancerre township itself, from Bué and Saint-Satur, and from the districts of Crésancy and Verdigny. Sometimes even better than these is the Sancerre made in Chavignol, a village celebrated for its 400-year-old elm tree in the main square, and for its tiny goat cheese called *crottins,* or "droppings," which can be aged for years and become as hard as stones. A bottle of Sancerre from the vineyards of Chavignol is easy to identify: the growers have so much pride in their wine that they insist that Chavignol occupy a choice place on the label, sometimes even in larger characters than the Sancerre *appellation* itself. Some excellent Sancerre rosé is made from the Pinot Noir grape.

Further down the Loire, near Orléans, wines produced at the time of Charles VII, in the 15th century, were termed "the best and most appropriate that one can find to suit the human body." However, today only a few light, unpretentious table wines for daily use are made there of Pinot Gris and Pinot Meunier. Below, just past the old city of Mehun-sur-Yèvre (where the young Charles received Joan of Arc in 1429), the district of Quincy makes an excellent white wine—but too little of it to be widely known. There are some 500 acres of vineyards, shared by some hundred growers, who prune the plants in the local "pistol" fashion, with the grape-bearing shoot cocked back on its vine prop two yards above the ground. Half of the yield is usually reserved to customers who regularly purchase it in the cask. A few bottles reach Paris. Those who have tried this pale white wine particularly praise its freshness, its fragrance, and its peculiar taste, which seems to have been developed especially to go with oysters. Nearby Reuilly also produces white wine, somewhat similar to Quincy, but with a less pungent flavor. The wine is pure Sauvignon Blanc. There is some rosé wine, "bottled at the moon of March," which growers sell only regretfully.

THE GARDEN OF FRANCE

Past the beautiful castle of Chenonceaux is the wine-growing province of Touraine, often called the "garden of France" because of its fertile green valleys. The province is also remarkable for its extraordinary wealth of castles. Almost each of them has given its name to a wine-growing district: Amboise, Azay-le-Rideau, Chinon, Langeais, Moncontour, Montrésor, Montlouis, Montrichard, the ruins of Rochecorbon, and Saint Aignan. Cellars here are cut out of a yellow sort of chalk, called tuff, which insures moderate and constant coolness. Balzac described the singular aspect of some of the winegrowers' houses: "In more than one place there exists three stories in a house, dug in the rock, and linked by perilous staircases cut from the stone itself . . . The chimney smoke rises between the shoots and branches of a vineyard . . ." Sometimes, the houses have an almost upside-down organization, with the cellar on top, the dwelling part below, and stables on the ground floor.

According to legend, pruning was invented two miles from Tours in the Marmoutiers abbey founded in the 4th century by St. Martin. It is said that the invention was made by a donkey, which was caught clipping the vineyards and soon was imitated by the wine-growing monks who noticed that the wine from these vines had improved.

The soil of clay and chalk is particularly suited to Chenin Blanc, or Pinot de la Loire, which yields one of the most famous *appellations contrôlées* of the Loire —the white wines of Vouvray. The city of Vouvray, whose motto is "I gladden hearts," has deep cellars cut in chalk where hundreds of revelers can gather dur-

A FEAST in honor of the French Army. Wine was always a popular subject of such 19th-century woodcuts. *Bibliothèque nationale, Paris.* (*Photo S. Hano*)

ing the festivities of the local wine-tasting fraternity, *La Confrérie de la Chantepleure.* (Chantepleure is the poetic name of a winecask spigot, which sings—*chante*—as it is being opened, then weeps—*pleure.*)

The white wines of Vouvray can be sweet or dry, with a taste of fresh grapes or of ripe quince, a subtle flavor of almond, chestnut, or acacia. Rabelais sang their praises as "taffeta wine." The wines reach fullness after a decade of ageing in the bottle. Many are sparkling: for instance, the Vouvray Mousseux, made in the Champagne fashion.

BOURGUEIL

One of France's numerous Benedictine abbeys settled down around the year 1000 in Bourgueil, a few miles downstream from Vouvray. The monks planted handsome vine arbors in the gardens of the abbey, and made some wine. It was so good that during the harvest they had to stand guard throughout the night lest local farmers make way with some of the yield. Two or three of the abbey growths, the Clos de l'Abbaye and the Grand Clos, are still among the best wines of Bourgueil today.

The best red wines of Bourgueil (like those of Chinon, a nearby castle whose huge ramparts dominate the Vienne River) are made in Saint-Nicolas-de-Bourgueil with the Cabernet Franc. This plant is believed to hale from Bordeaux, but goes here under the name of "Breton." At Saint-Nicolas grapes are often torn away from stalks right in the vineyards with a wooden palette, and are thrown into a sort of lattice wicker funnel. The wine is made in conical, open wooden vats of no more than 1,000 gallons each.

Bourgueil wines, with a slight taste of raspberry, are reminiscent of a light Médoc; the red wines of Chinon are smoother, with a deeper color, and a bouquet —some claim—of violets. Bourgueil, according to French author Jules Romains, is a "wine for intellectuals: not loaded with tannin like Bordeaux wine, nor delightfully toxic like Burgundy."

A rare but excellent wine comes from the steep slopes along the Loir River, one of the secondary confluents of the Loire. Here the white wines from the Pinot grape have freshness, a handsome color of burned topaz, and sometimes a light sparkle. The red wines from the district have the reputation of keeping almost indefinitely—or, at least for fifty years—without losing their vigor and their spirit. Henry IV rated Loir wines with those of his beloved Jurançon.

Further downstream along the Loire, there rise the elegent bell turrets and

THE MYSTICAL GRAPE HARVEST. Apocalypse tapestry, 14th century. *Musée des Tapisseries, Angers. (Photo Giraudon)*

handsome towers of what once was the 10th-century fortress of Thibaut the Cheater, Count of Blois and of Touraine. It is Saumur, the home of the renowned military equestrian school, *the Cadre Noir,* which holds a *carrousel* (a parade with demonstrations of horsemanship) every July 29. At the end of the *carrousel,* the officers traditionally open bottles of sparkling Saumur by neatly decapitating them with their sabres. Local producers are grouped in Saint-Hilaire-Saint-Florent, a small village stretching along the Loire, under which there is nearly a mile of labyrinths dug in the tuff, with a tiny railroad track, no longer used today. There are two types of sparkling Saumur wines: the Pétillant, with a slight sparkle (this wine is occasionally available in the United States) and the Saumur Mousseux, a fully sparkling wine. Not far away, in Champigny, one of the region's best red wines is made from the Cabernet Franc plant. It is somewhat similar to the wines of Bourgueil and Chinon, but more full-bodied.

ROSE D'ANJOU

Rosé wines from Anjou, chiefly produced along the Layon river, are often called Rosés de Cabernet, the plant from which they are made. The annual production exceeds three million gallons, and the best of these wines have a rather deep color and aroma, combined with lightness. The authentic Rosés d'Anjou are perfect thirst-quenchers when young. They also age without losing vitality, their color turning to golden-green or dark old gold, while developing a generous bouquet. There are, however, many imitations, usually easily identified by being almost colorless, having a taste of sulphur, or being excessively sweet. (Sweetness is sometimes deliberate in fraudulent wines because it is known that *le doux fait passer les autres goûts*—"sweetness overwhelms other tastes.")

Some ten miles southwest of the city of Angers, the charming village of Savennières has a fine white wine produced from vineyards along the river, and bearing the *appellation* of Côteaux de la Loire. The best vineyards are those of the Roche-au-Moines, producing a dry and crisp wine; facing the Roche-au-Moines are the slopes of the Coulée-de-Serrant, whose wine is liquorous and often excellent, although chiefly of academic interest—so little of it is made that hardly any ever even reaches the French capital.

Much wine is drunk by the Angevins, the inhabitants of the province of Anjou. The local saying is *"Angevin, sac à vin"* (wine-bag), and this gave its name to the local wine-tasting fraternity, the *Confrérie des Sacavins*.

MUSCADET

The most widely known wine produced along the Loire is doubtless Muscadet, the gay, often slightly acid, fresh white wine which is the traditional fare and fuel of French congressmen debating the affairs of state in cafés around the *Chambre des Deputés,* the French House of Representatives in Paris.

The wine is made along the Loire before it reaches the ocean in the city of Nantes, from the Melon grape plant, called also Muscadet. The plant was introduced into Brittany in 1709, after the destruction of the then existing vineyards by a winter so cold that the ocean along the coast was frozen solid. The people of Nantes, who used to pay rent with wine, promptly set out new vineyards, and their wine today has conquered France and is rapidly captivating the rest of the world. A wine most widely used with seafood, it is often plain and without great distinction. But it is also often superb, with a pale yellow color, dry and supple, sometimes extremely dry.

Muscadet is best when the month of October has been dry and sunny, and when enough rain has fallen in the earlier part of the year; such were the years 1953, 1959, and 1961, among the best recent vintages. Muscadet has nuances that vary from owner to owner and from cask to cask. If drunk too young, the wine can be too sharp, and as a rule the most distinguished wine is not bottled before Easter. Excellent Muscadet comes from the region of Sèvres-et-Maine, which is indicated on the label, and from the Côteaux de la Loire. The plain *appellation contrôlée* Muscadet is unpredictable—sometimes good, but sometimes too soft and lacking character.

Wine growing here is usually a family affair, and small vineyards predominate. Many burghers of Nantes own a few plants around the city, and farmers who raise cattle or grow other crops like to tend a small plot of vines, chiefly for their own use. Wines from different parcels are generally blended, and the must is left to ferment on top of a thin layer of lees, in an oak cask into which a sulphur wick has been dipped. Wine is racked into glass-lined vats, and, after fining, is bottled. Much Muscadet is bottled in the month of March, in order to be ready for the wine market and the festivities that take place throughout the region.

The popularity of Muscadet in France itself is fairly recent, having rapidly risen only after World War I. The total production has not exceeded five million gallons a year, and it should be mentioned that some wine sold on the counters of Parisian cafes under the name of Muscadet should be drunk with skepticism.

Another wine in the same region is believed to issue from the descendants of a few vines that survived the great freeze of 1709. It is called Gros Plant, after the name of the grape. Less sharp and less lively than the best Muscadets, Gros Plant has smoothness and finesse and a pale color whose translucency is enhanced by Alsatian-type thin tall bottles. Gros Plant is slowly becoming known, but the demand far exceeds the limited supply.

TREADING GRAPES, miniature.
Bibliothèque Sainte-Geneviève, Paris. (Photo Etienne Hubert)

RHONE, PROVENCE AND MIDI

From a glacier in Switzerland to the Camargue (the French "Far West," west of Marseilles), the Rhone River follows vineyards. Beginning in the Swiss Valais, where wines are cool and have a sparkle, it goes on to the French Savoie, with its light and pale growths; farther south, past Lyon, both the Rhone and its wines acquire a fuller body—reaching maturity along a stretch of some 50 miles between Vienne (called "the vinous") and Avignon. These are the famous Côtes du Rhône, with vineyards steeply rising above the Rhone Valley, producing distinguished and potent red wines. In the French southeast, the wines are sunnier and less serious; they are the jolly *rosés de Provence,* wearing not a robe but a slim pink bikini, and are best suited to be drunk on a sunny beach.

THE RHONE VALLEY

In the mountainous north, the region of Savoie (it used to be Sapaudia, "land of pines") produces four or five million gallons of wine annually. The best of the wines is usually the Roussette de Seyssel, from vineyards which Carthusian friars of Arvières planted in the 14th century, and which have been nursed since with religious devotion by Augustine, Benedictine, and Capuchin brothers. The plant is the Roussette Haute or Altesse, which came from Cyprus, yielding a golden, tender, supple, full-bodied and yet delicate wine. Some Roussette is made into sparkling wine, enjoyed locally. A few rosé and red wines (notably from, the Mondeuse, a red wine grape) forecast more serious business as the Rhone comes down the mountains and approaches Lyon. In Lyon, the river makes a turn and flows due south, through one of France's oldest wine-growing regions, built up by Roman conquerors around the town of Vienne. Here the first years of the Christian era witnessed the domestication of *Allobrogica,* a plant whose dark red berries can ripen in spite of the cold and frosts. *Allobrogica* wines, to which resin was sometimes added, were popular with the Romans, and traveled both to Italy and to the colonies in Britain aboard wine boats and on muleback. The once flourishing wine industry has left so many ancient potsherds that they are used as fills under modern constructions. Today this region still produces wine— about ten million gallons a year, which is more than the output of the Côte d'Or of Burgundy; but these wines have not been found worthy of being granted their own *appellation contrôlée.* Much of the wine today is made from the Syrah (or Sérine) wine plant, believed to have been imported from Shiraz in Persia by Roman soldiers. These red wines have a solid body, much tannin, and a peculiar inky taste which some do not dislike.

A few miles to the south, the Côtes du Rhône begin. Côte Rôtie wine, from the village of Ampuis seven miles south of Vienne, has a beautiful translucent red color, and a fine bouquet. It is not easily obtained, and neither is Château Grillet, from the smallest vineyard in France to have its own *appellation contrôlée.* On the river's west bank, Château Grillet produces a rich yet dry white wine, somewhat reminiscent of a good Hock, yet with a unique spicy flavor derived from the Viognier vine plant. Condrieu, nearby, is a venerable vineyard, also

VINEYARDS OF TAIN-L'HERMITAGE, with the Rhone in the background. On this slope a good eye and sure feet are requisites. *(Photo Chapman)*

producing white wine from the Viognier plant. A fortified dependency of the archbishop of Lyon in the 12th century, it produces mainly a dry wine, yellow and with a strong bouquet of its own. Condrieu wine can be slightly sweet, and also sparkling if measures are taken to slow down fermentation in the vat.

Further south is the small town of Tain, with vineyards of the Hermitage (or Ermitage) and Crozes Hermitage side by side on the left bank. Historians disagree about who was the hermit entitled, in the town's account of 1598, to "four *écus* to purchase him a robe," but some claim it may have been St. Patrick. In fact, the vineyard existed in Roman times, and no one knows just why it is called the Hermitage. Before phylloxera devastated the vineyards, these wines used to rank with the best Burgundies and Clarets—and Hermitage wine went to Bordeaux to strengthen local wines, reaching England in this devious way.

The true Hermitage wine is made of a mixed harvest from variously exposed terraces facing south and rising, in ladder fashion, above the river—unsparing of man's labor since the slopes are inaccessible to machinery. This single vineyard slope is two miles long and half a mile in width; it is divided into eighteen sections, and yields some 70,000 gallons of wine a year. Almost one third is white wine from Roussanne and Marsanne grapes, and two thirds red from the Syrah. The white is full-bodied and dry, but the red is more interesting, slightly harsh and even bitter in its youth, then growing smoother. With a lively color, it is very long-lived and full of fire. The adjacent Crozes Hermitage vineyards have somewhat similar wines but lighter—and about twice as abundant.

Four or five miles downriver on the Rhone's west bank, the wines of Cornas, also made from the Syrah grape, are sturdy but less distinguished. Saint-Péray, nearby, has a lively and generous white wine, rich in alcohol, perfumed, but subject to maderization (oxidation and yellowing into Madeira-like wines) because of its long stay in the cask. It comes from the Roussanne and Marsanne grapes, precociously matured by summer heat on sun-drenched slopes.

East from the Rhone, along the river Drôme, the little town of Die has been famed for its wines since Roman times. Clairette de Die is the reflection of a climate which is at the same time Alpine and Mediterranean. With a pale yellow golden color and a strong muscat taste, the wine results from a blending of Clairette and Muscat grapes growing on stony slopes along the Drôme. The wine is made to sparkle by an old rustic method which takes advantage of the cold of the winter and uses several successive finings and filterings to slow down fermentation, so that the wine can "go to work" again in the spring and make its foam.

CHATEAUNEUF-DU-PAPE

Finally, in the southern part of the Côtes du Rhône, is the celebrated district of Châteauneuf-du-Pape, once the summer home of the Avignon popes. The Châteauneuf ("new castle") was first built in the 12th century as a fortress of the Knights Templar, which became the property of Frederick Barbarossa, who gave it to the bishops of Avignon. One of the bishops, who later became Pope John XXII, restored and enlarged the austere Templar building to make it a summer residence, The popes had vineyards planted, which changed hands many times until the 16th century, when the Châteauneuf was reduced to ruins by the Baron des Adrets, leader of the Protestant Huguenots (the *coupe de grâce* was given by a German bombing raid in 1944). Today, only the fortified facade, dominating the cypresses, olive trees and vineyards, bespeaks the past splendor when popes, and later the schismatic popes, dwelt here. The red wine of Châteauneuf-du-Pape probably owes its reputation to an assemblage of thirteen different plants. Grenache is strong in alcohol and gives the wine much warmth and smoothness; comprising about 40 percent of the wine, it is mixed chiefly with Syrah (30 percent), which gives the wine body, ageing potential, and a deep color. The two are joined in the vat by Cinsault, Mourvèdre, Caunoise and Vaccarèse, which give wine suppleness, freshness, and bouquet. These represent about 20 percent in the classic formula, and the remaining 10 percent is chiefly made of two white plants, Clairette and Bourboulenc, which contribute fire and brilliance. The formula varies from one winegrower to another.

Fermentation goes on for at least two weeks and often longer. Only during very warm years are grapes removed from stalks before the pressing, and then the wine's tannin content must be increased by preparing a special must, made from grapes with their stalks left on, that is mixed with the others during fermentation. To increase the bouquet, some wine makers resort to "carbonic maceration": whole bunches are placed in small, closed cement vats of about 700 gallons, and under the weight of the grapes some juice is made and partial fermentation starts, releasing carbon dioxide and building up the pressure in the vat. After eight days or so of this process, during which grapes exhale a wealth of essence and perfumes, pressure is gradually released, the pulp is pressed and the juice pumped into other wine vats, where fermentation will go on until completed. The wine matures in oakwood casks.

The alcoholic content is high: about 14 percent, highest of any of the great French red wines (the required minimum for Châteauneuf-du-Pape is 12½ percent). The wine is full-bodied, of a dark crimson color, warm and a bit harsh in its youth; it reaches smooth fullness only after three or more years of ageing.

The white wines of Châteauneuf-du-Pape are subjected to a minimum of treatment during vinification. They are bottled as late as possible, and are very long-lived. Although rich and flavorful, they are no equal to the reds. The total annual production of Châteauneuf-du-Pape usually exceeds one million gallons, of which only about 2 percent is white wine.

Tavel

Across the river from Châteauneuf-du-Pape, the village of Tavel produces one of the world's best rosé wines—some say *the* best. Praised by Honoré de Balzac, it is one of the few rosés which age well, retaining for several years its fruitiness and flavor. Famed since the 17th century, it was lyricized by the poet Ronsard as being "sunshine in flagons." Of a beautiful ruby color, brilliant, full-bodied, heady, it is very dry and yet supple. The annual production is about 400,000 gallons, and its closest rival is the Lirac rosé from a nearby village.

Downriver from these aristocrats, the balance of the southern Côtes du Rhône spreads over 80 townships. Most of the wines are made in cooperatives; many are fair, but none outstanding.

PROVENCE

To the east of the Rhone estuary, the vineyards near the river Var may yield up to fifty million gallons of wine a year. Here the best wines are produced along the coast, above the rocks that plunge into the sea, and around colorful and hospitable villages, shaded by pine trees and mimosas. The rosé wines are fresh and fruity, without pretensions, most pleasant when young and drunk with sea urchins, shellfish, a local fish called *loup* (whose delicate flavor is enhanced by a pyre of fennel) or with grilled sardines.

The inland part of this region, the middle and upper Var, is the kingdom of the Côtes de Provence, producing wines of *appellation contrôlée* and *vins délimités de qualité supérieure* (V.D.Q.S.) which can be drunk today nearly everywhere in the world. They are at their best along the wine road leading east of Toulon through the most handsome vineyards of the region. It is a route that passes old farmhouses, haughty ruins and, perched on their stony hills, villages nestled around their fountain and shaded by plane trees. The silence of the sun in the luxuriant

CHATEAU SAINTE-ROSELINE, Le Muy, Var. *(Photo Spirale)*

fragrant growths is interrupted by the endless squeaking of cicadas, and by all sorts of winds: the humid *gregali* from the northwest, the *tramontane* or *montanière* from the north, the potent and dry *mistraou* from the northwest, the cool *pounen* which breathes from the west, and the winds from the sea—the *miegiou* from the south, the *levant* from the east, the *eisseroq* from the southeast and the *labech* from the southwest, carrier of rain.

Pierrefeu, Gonfaron, Vidauban, Les Arcs, Tradeau and Lorgues are the main wine districts inland. Overlooking the sea are the colorful and shaded villages of Ramatuelle and Croix Valmer, Gassin—on a hilltop that would like to be inaccessible—and, down below, the fishing village named after a holy man but which, some believe, now belongs to the Devil: Saint-Tropez. This region is still part of the kingdom of rosé, fruity and tasty, with a silky, transparent dress, admirably suited for bouillabaisse and other dishes made with fish.

The rosés of Provence gained fame in the early 1930's, thanks to quality rulings by the local growers' syndicate, which set up severe standards. The wine most fashionable in Saint-Tropez in summer is a rosé made just a few miles away, below Gassin—the fresh and dry Château de Minuty, which comes in a club-shaped bottle. Less wealthy tourists buy the wines of Bastide Blanche, which is also the name of a beautiful beach. The rosé is quite strong and reddish and often drunk with ice, and the white is, in fact, almost rosé, with its pink tinge more or less pronounced, depending on the year. Some excellent rosés are made by the Domaines Ott, which own vineyards from Bandol to Draguignan, and also produce red and white wines. Like their big brothers to the north, these wines have their fraternity, the *Confrérie des Chevaliers de Méduse,* founded in Marseilles in 1690. On St. Vincent's day, January 22, commanders of this Bacchic order gather in one of the villages whose wine has reached a peak in reputation. They attend mass, then sit around to taste local wines and to have an abundant lunch, shared by the neighboring growers.

Further east, toward the town of Nice where Englishmen invented the Côte d'Azur a century ago, near Grasse, Vence and Antibes, there are a few delicate red wines. The best are made around the town of Bellet from local plants, La Folle and Le Braquet.

FRANCE'S LARGEST VINEYARDS

From the Rhone west to the Pyrenees along the Mediterranean coast and north to the foothills of the Massif Central is the world's largest vineyard—although it is far from being one of the best. Uninterrupted, the vineyards of the Midi cover more than a million acres—5 percent of the world's vineyard acreage and almost one third of the vine-covered land in France. The region produces more than 40 percent of the volume of French wines, and is the principal means of livelihood for one and a half million people.

In answer to the pressing demand for wine, the regions of Languedoc and Roussillon chiefly produce abundant, ordinary, cheap wines. The requisite here

BACK TO THE FARM
after a day's harvesting.
(Photo Ribière)

Right: This poster announces
the greatest upheaval that ever hit
the international wine business.

is a high alcoholic content, on which the price of the wine is based. The vine plants are usually of high yield and low quality, particularly Aramon and Carignan. When the vine louse phylloxera reached France from America in the 19th century, it took only a few years to ruin the vineyards in this region. Salvation came from the city of Montpellier, where Jules Planchon, a botanist and professor of pharmacy, and Gaston Bazille, a landowner, made the first grafts of European plants on phylloxera-resistant American root stock. Then a nursery owner called Louis Bouschet started developing hybrid plants, by crossing and selecting existing varieties. Many hybrids now bear his name. Most of these vines give black, strong wine, without distinction. For a time hybrid plants contributed to the salvation of the wine industry, but now the planting of hybrids is restricted to the best ones only.

After the solution to the phylloxera problem had been found, winegrowers frantically replanted their vineyards, whose loss and reconstitution cost them a total of about a trillion "old francs"—nearly two billion dollars. Meanwhile the shortage of wine had opened the French market to wines from Italy and Spain, and from Algeria, where new mass production vineyards were rapidly spreading. When phylloxera reached these competitors early in the century, the means of beating the pest were already at hand, having been developed in Languedoc.

Today, the quality of wines from Languedoc has improved somewhat, though the wine is seldom more than *gros rouge ordinaire,* the plain red stuff about which it is said that *il tache et pousse au crime*—"it stains and incites to crime." But planting is now more discriminate, equipment modern, and techniques well studied. Some brand-name wines have achieved considerable success—the Vins du Postillon, for instance, which go directly from the producer to the grocer, or wines from the large firms, such as the Etablissement Nicolas, which sells wines, aperitifs, and hard liquor, and whose avowed aim is to make it unnecessary for the average consumer to have a wine cellar of his own. Nicolas ships its wines in bottles and delivers it to the home, totalling up sales of as many as 400,000 bottles a day. Two-storied cellars in the suburbs of Paris hold some 10 million bottles, and the winehouse has a quarter of a million gallons of wine in blending vats, more than 300 stores in Paris, and more than 400 in the rest of France—producing a gross annual income, in terms of dollars, written in nine figures. Such huge winehouses must be supplied by huge wineries, which are of the cooperative type, and have been called "the cathedrals of the Mediterranean Midi." The largest winery, traditional supplier of Nicolas, is in Marsillargues, halfway between Nîmes and Montpellier. It has about 700 growers as its members, and its vats hold more than two million gallons at a time.

There are a few wines here that emerge from anonymity. Banyuls, for instance, is a wine of *appellation contrôlée* made with grapes dried on the plant, with an alcoholic content reaching 18 to 22 percent (after addition of alcohol and ageing for three years). The Muscat wines of Frontignan, near Sète, and of Lunel, between Nîmes and Montpellier, can also be quite palatable when taken in small doses. Among the best plain table wines is the Corbières, a V.D.Q.S. wine produced along the Mediterranean coast and in the eastern Pyrénées; the annual output is upward of 60 million gallons a year. Though inexpensive, it is a good notch above ordinary, nameless wines. Corbières can be red, rosé or white, but the best without doubt is the red, issued from the Grenache, Carignan, Terret and Cinsault grapes.

SPRINGTIME IN THE VAR. *(Photo Ribière)*

THE FORTIFIED CHURCH OF HUNAWIHR. *(Photo Franval)*

ALSACE

The wealthy province of Alsace has been fought over time and again by rival Germany and France, and its inhabitants have kept something from both. A case in point is Alsatian cooking, which seems to borrow from both countries, and in any respect is superb. There are so many different types of sauerkraut, for example, that it is difficult to choose among them, and the temptation for the visitor is to stay and try them all.

Geese figure prominently—and deliciously—in Alsatian cooking, as do hare, deer and boar. But before the main dishes are served, there will be such delicacies as cheese or onion pie, a *brioche*—fluffy cake—stuffed with *foie gras,* woodcock pie, frog's legs in cream or smoked trout with horseradish. To round out the meal, the Alsatian host may offer such cheese as pungent Munster or creamy Géromé, and then perhaps a tart flaming with kirsch, or a Kugelhopf yeast cake with raisins soaked in rum or kirsch, the likes of which cannot be found elsewhere. And of course there will be a rich array of Alsatian wines to accompany the courses.

Although similar to the wines of the Rhine valley, Alsatian wines do not reach the same towering heights that true Rhine wines can attain—and as a matter of fact, the vineyards, while they parallel the Rhine, do not actually accompany it. Instead they run between the rich and fertile plain of the Rhine's confluent, the Ill (from which Illsass, thence Alsace), and the pines of the Vosges Mountains.

Taking advantage of the long years of conflict between the French and the Germans, Alsatian winegrowers long managed to avoid the obligations—sacrosanct elsewhere in France—of the governmental controls of *appellations contrôlées d'origine.* This independence (some would say anarchy) ended on October 3, 1962, but in fact the new regulations pretty much followed the traditional Alsatian usage. Thus, instead of calling their wines by their place of origin, most Alsatian wines continue to be identified by grape variety, although there has been added the denomination of *appellation vin d'Alsace contrôlée*.

The emphasis in Alsace is on white wines, and here the choice is wide. There are five grape varieties that qualify for the adjective of "noble": Riesling, Gewürztraminer, Muscat, Pinot Blanc, and Tokay. A sixth, Sylvaner, is often given a title of nobility in France, but not across the border in Germany.

From a practical viewpoint, the noblest plant in Alsace is the Riesling, which probably originated in Germany, where it is believed to have existed in Roman times. This king of Alsatian wines reigns with delicacy and a discreet bouquet. Most Riesling grapes are gathered late—at the end of October—but the best grapes are left to mature well into November, to yield *Auslese*, or "special selection," wines. At the same time both dry and fruity, Riesling is equally as good with oysters as with *foie gras*, trout and sauerkraut.

The Gewürztraminer, which some wine lovers place above the Riesling, has a taste that is unmistakable. Full-bodied, dry or only slightly sweet, it is one of the most intensely aromatic and spicy of wines anywhere, and goes well with such dishes as cheeses and sweets.

Muscat, with its pronounced taste of fresh grape, develops a peculiar savor which makes it an interesting dry aperitif. Today, Muscat is almost always made of a lesser variety, the Ottonel, which cannot stand comparision with a true white Muscat. The Pinot Blanc, also called Klevner, has great freshness, due to a slight acidity. It is a balanced wine, delicate and mellow. The Tokay of Alsace, not related to Hungarian Tokay, goes into an excellent white wine particularly suited to *foie gras*.

The semi-noble Sylvaner is a great favorite throughout Alsace, as it is more productive than the five greats. Cool, light and fruity (sometimes almost sparkling), it goes well with mussels and fish or various delicatessen foods. Another favorite is the Zwicker—not noble and, in fact, not a vine plant at all, but a blend of the Chasselas grape with local varieties such as the Knipperle and the Goldriesling (not to be mistaken for the true Riesling, to which it is not related). Bottled at the end of the winter, Zwicker wines may contain small amounts of carbon dioxide, which give them a frizzly sparkle, known as *moustille*. Edelzwicker (from *edel*, noble) is not a plant variety either, but a blend of several more or less noble plants, often grown in the same vineyard and picked together.

THE ALSATIAN WINE ROAD

La route des vins, or wine road, of Alsace, begins at Thann, just west of Mulhouse, and runs northward, almost in a straight line, to Strasbourg, passing first through the *département* of Haut-Rhin (the southern part of Alsace) and then into that of Bas-Rhin (the northern part). The visitor proceeds between the forest

HARVESTING BUCKETS AT RIQUEWIHR. *(Photo Chapman)*

GRAPE HARVEST AT RIQUEWIHR. *(Photo Chapman)*

to the left and the plain to the right, among ruined castles, cathedrals, and some of the prettiest villages in France. Thann has its famous Rangen growth, so heady that it has given rise to a mild local curse, "Let the Rangen shock you!" A few miles to the north, Guebwiller has four well-known vineyards facing due south—Kessler, Kutterle, Searing and Wann. Rouffach, once the property of the jolly King Dagobert (who, in a popular song, wears his pants inside out) is dominated by the Tower of the Witches, and belongs to the same district.

Near Colmar, the village of Eguisheim makes excellent wines, and its ancient cellars and houses, kept intact since the 16th century, attract many visitors.

A visit to Colmar, the capital of Alsatian wines, is well worth while. Once the Roman *Villa Columbaria*—the domain of the doves—and later one of the imperial residences of Charlemagne, Colmar is the home of the local Viniculture Institute, and of nearly all the wine merchants in Alsace. It is also the meeting place of the *Confrérie Saint-Etienne d'Alsace,* a wine fraternity founded in Ammerschwihr 600 years ago, whose members call themselves Gourmets and for centuries have tasted all wines put on the market.

The town has kept all of its timber framings, its taverns with low ceilings, its tall, sharp, steep roofs with small dormer windows, its flower-covered balconies, its ancient porches and Gothic churches. Almost too pretty to be true, it is a postcard town, like Venice or Copenhagen. Alsatians love their past, preserve it carefully, and a walk through the city of Colmar may leave the strange impression that, like the Baron Munchhausen, you have been riding a time machine into the past.

After Colmar comes the fortified town of Turckheim, whose most famed growth, the *Brand,* draws its vigor from the sandstone of its slopes. Aside from a range of white wines, Turckheim produces one of the few Alsatian reds, made from Pinot Noir and called "Blood of the Turks."

THE FOUNTAIN OF WINE

The wine road now leads to Beblenheim, Mittelwihr (so sunny that almonds ripen here), and, above all, to Riquewihr, one of the joys of Alsace. Medieval Riquewihr is perfectly preserved, complete with torture chambers and arrow slits in fortified walls. All the surrounding vineyards, from Sigolsheim to Ribeauvillé, through Zellenberg and Hunawihr, produce excellent Riesling. And if you come to Ribeauvillé on September 1, you'll quench your thirst for free: the fountain of wine flows on the town's main square as popular festivities commemorate the "Day of the Piper." This is a festivity dating back to the Middle Ages, marking the day when, traditionally, the powerful musicians' corporation gathered to pay harmonious respects to the *seigneur*. Altogether, the vineyards of Haut-Rhin produce about a million gallons of wine a year, including most of the Alsatian wines of *appellation contrôlée*.

Through forests and vineyards, past the hop fields and the tobacco patches of the picturesque Hohwald, the road now leads into the Bas-Rhin, producer of about half a million gallons of wine a year.

The vineyard stretches from Ribeauvillé to Heiligenstein, and starts again at Obernai and at Molsheim. This is light soil upon which wines from Sylvaner and Gewürztraminer have built their reputation. The town of Barr and vineyards around it are in the center of vinous activity, producing more than 60 percent of the area's output. The July wine festival lasts a week. Above, on the hilltop, the ruins of the castle of Andlau stand as a monument to the sad story of the Empress Richarde, whose husband accused her of being unfaithful, and made her cross a fire, barefooted and wearing only a nightgown. The empress retired to an abbey she founded nearby. The terrace of the castle is an excellent vantage point: from here can be seen the Alsatian plain stretching out, bordered in the distance by the Black Forest; and, just below, slopes covered with the noble Riesling.

RIQUEWIHR STREET SCENE during harvest time. *(Photo Chapman)*

LORRAINE

West of Alsace, the province of Lorraine once had abundant wines, but the vineyards have shrunk. There are now only memories of the red wines of Pinot de Bar and of Gamay, once glories of the bishoprics of Metz, Toul and Nancy. Lorraine today still has a few pleasant *vins du pays,* the most interesting being the *vin gris* from the upper Moselle valley between Metz and Toul. Made from mixed red and white grapes, this wine has a lusterless pale pink color. Pleasant, light rosés are also made in and around the villages of Bruley and Liverdun.

JURANÇON

In 1553, when Henri de Navarre, the future *Bon Roy* Henri IV was baptized, he was given a taste of wine—not Bordeaux wine, but Jurançon, from the heart of his own Navarre further south on the Spanish border. This is the wine about which Colette wrote, "When still adolescent, I made the acquaintance of a prince, full of fire, commanding and treacherous, like all the great seducers: the Jurançon." Today, alas, Jurançon wine is little known. It deserves a better fate.

One of the three wine-growing districts of the Pyrenées, Jurançon, is almost hidden among sunny hillocks, rich, green pastures and cornfields, chestnut trees and fragrant acacias. The vineyards, under the pure, somewhat rarefied atmosphere below the snowy mountain peaks, produce a small amount of excellent white wine from local plants that are not found anywhere else—Petit Manseng, Gros Manseng, and Courbu. The production seldom exceeds 400,000 gallons of genuine white Jurançon a year, and a few gallons of red wine.

The *blanc de blanc brut* in Jurançon is made in a manner particular to the region: the grapes are thrown into huge oak vats, and the juice drips down through the lattice underneath. Only the result of this first drainage is used to make the quality white wines; wine from the following pressings, though very palatable indeed, is considered of secondary quality by the local quality control association, the *Viguérie royale du Jurançon,* which issues its label every year only after examining and tasting the wines. Today, a few wine lovers in Europe and in the United States can obtain this wine, and there is hope that this trickle will increase. Local gourmets, however, do not complain; they are happy to drink most of the Jurançon *blancs* themselves as a favorite fare with the trout caught in their mountain brooks, or with the *buissons d'écrevisses* (literally, the "bushes of crayfish") gathered in the smaller streams.

Drunk the year after it is made, Jurançon *blanc de blanc* has its own inimitable freshness and fruitiness, often attributed to the mountain air breathed by the vines. Kept in the bottle a few years, its character changes entirely—many prefer the dignified, balanced qualities of aged Jurançon to its youthful recklessness.

Another Jurançon wine, made from slightly overripe grapes is entirely different. Its alcoholic content may exceed 13 percent, hence its fermentation slows down and sometimes stops entirely during the winter, to awaken again with the spring. Sweet and smooth wines of this kind, less delicate than the sweet Sau-

ENAMELED TILE, 18th century. *Musée des Arts décoratifs, Paris. (Photo Chapman)*

ternes, are excellent as aperitifs. They are just as appropriate with *foie gras,* or duck liver prepared with *morilles* (a local mushroom), cheese, or dessert.

Nearby, just north of the cities of Tarbes and Pau, another little-known wine will pleasantly surprise the visitor: it is the red wine of Madiran, practically unknown outside the region. Madiran wines can have the potency and race of some of the great reds. They age well, and keep for many years a pleasant ruggedness, inherited perhaps from the harsh winters, stormy autumns, and lead-hot summers of the Pyrenean foothills.

East of Bordeaux, in the *département* of the Dordogne, there are two more wine-growing areas, Bergerac (the land of Cyrano) and Monbazillac. Bergerac has a wide range of white wines, from dry to sweet, unpredictable and of little consequence and also makes some red wines. Monbazillac is a soft, rather ordinary sweet white wine which can, occasionally, have a pleasant Muscat aroma. The area around Cahors, once famous for its reds but wiped out by the phylloxera, is having a renaissance.

Northeast of Toulouse, the vineyards of Gaillac produce a wine from a local plant, the Mauzac; it is usually sold anonymously as *vin de comptoir,* and is served in bars and cafes to customers who simply ask for *un petit blanc.* Some Gaillac is made into sparkling wine. The standard champagne method is used. The best of such wines are made with a local method, *à la Gaillacoise,* with part of the wine's original fermentation taking place in the bottle.

ARBOIS AND JURA

The wines of the Jura, *département* bordering Switzerland in the old province of Franche-Comté, were once much favored by the courts of Burgundy and France. And it was at Arbois, on the slopes of the Jura mountains, that Louis Pasteur began his researches into alcoholic fermentation.

This region of pine trees, reddish forests and pastures is dotted with picturesque towns climbing up mountains or settled atop hills: Poligny du Jura, l'Etoile, Château-Châlon, and Arbois itself, which gave its name to the *appellation* that covers five types of wines: red, rosé, white, and two more, found only in Franche-Comté: *vin jaune,* or "yellow wine," and *vin de paille,* "straw wine."

Vin jaune is made only with the Savagnin plant, which Benedictine monks of the abbey of Château-Châlon are said to have imported from Hungary around the year 1000. The grapes are harvested at the beginning of the winter, after the first frosts, and hung in well-heated and aerated attics, to be crushed toward the end of February. After the usual rackings during the first year, this "frost wine" is stored in barrels. When the weather grows warmer, there appears on the surface a thin film of that same *flor* yeast which is so important to Sherry wines. Six years later—no less, the wine has acquired a bright yellow color, and a taste reminiscent of Sherry and walnut. (The wine, in fact, has a few similarities to unfortified Sherry, though it has less alcohol: 11.5 percent for Arbois and l'Etoile, 12 percent for Château Châlon, 11 percent for Côtes du Jura.) The "yellow wines" are entitled to a special, stumpy bottle, the *clavelin,* which holds about 20 ounces.

The second Jura specialty, *vin de paille* or straw wine, is found today only in the district of Voiteur. It is made by letting grapes dry on stray mats until they are almost raisins. The pressing yields a highly concentrated must that ferments slowly. Such wine must age eight to ten years before it can be bottled—and further ageing in the bottle gives it a subtle flavor. It has at least 14 or 15 percent alcohol, combined with a high content of unfermented natural sugar.

Next to these rare beverages, of which only a few thousand gallons are made every year, the Jura has some simpler wines—white wines from Savagnin, Melon and Chardonnay vines, fruity *rosé* wines, and sparkling wines made in the classical Champagne fashion. Small amounts of red wine are made from local plants, the Trousseau and Poulsard, which once knew glory at the French court.

The wines of the Jura were praised by Henri IV and Rabelais, but the vineyards have suffered from wars and epidemics, from the phylloxera and the planting of mediocre stock. The yield is now under a million gallons—only a tenth of what it used to be.

BARREL OF A 16th-CENTURY MUSKET, with vine decorations.
Musée des Arts décoratifs, Paris. (Photo Chapman)

CORSICA

Off the southeast shore of France, and visible from Nice on a clear day, is the island of Corsica, Napoleon's birthplace. It is a hospitable island where flowers bloom all year round, and where the smallest village perched in the mountains produces its own wine.

French ways of making wine have largely been adopted in this island, but the vine plants remain chiefly Italian: Genovese, Montanaccio, Sciacarello, Canaiolo Nero, Aleatico, Moscato, or Biancolella. The town of Sartène, in southern Corsica, produces the most wine but not the best. The wines with most distinction come from the north, from Cap Corse, the fingerlike promontory pointing at France. They are labeled as *vins du Cap Corse,* with a few particularly good growths named for the villages that produce them. Outstanding among these is Patrimonio.

The finest growths of Corsica are chiefly rosé and red, with a character of their own. Though not excessively strong, they are renowned for being treacherous, mainly because they are so easy to drink in large amounts. And unlike certain Beaujolais wines, they are remorseless—sharing, in that respect, the character of Corsicans, who have been known to pursue a vendetta for several generations.

Generally, the red wines are most interesting in this rocky island. Hot summers can give them a high alcoholic content, and the best red wines are heady and with a strong perfume. One of the most interesting growths is from the district of Tallano. Dark red, it becomes clearer after four or five years in the bottle, and after 15 years may acquire a beautiful clear ruby color, without losing any of its bouquet.

Few of the wines, except for some ordinary rosés marketed in France, are known outside of Corsica, and some are not even known outside of the villages where they are made. The annual production has been increasing, having reached some five million gallons, and there is a possibility that these wines may become available on the world market, should this trend continue.

Following pages: THE VINEYARDS OF BERNKASTEL on
the banks of the Moselle. *(Photo Bildarchiv-Irmer)*

Western Europe

ITALY
GERMANY
SWITZERLAND
AUSTRIA
SPAIN
PORTUGAL
BELGIUM AND LUXEMBOURG

WINEGROWERS' FESTIVAL AT VEVEY,
Vaud. Popular engraving. *(Photo Etienne Hubert)*

ITALY

Wherever a traveler sets foot on Italian soil, he is right in—or at least never far from—a wine district. Italy—ancient Oenotria, "the land of wine"—is barely larger than half of France and smaller than California, but it produces in the neighborhood of a billion and a half gallons of wine a year. This is nearly as much as—and sometimes more than—its neighbor across the Alps, and six or seven times as much as production in the United States. Italians also drink large quantities of wine: the annual per capita consumption is around 30 gallons.

In Italy vines grow up trees, in cornfields and artichoke gardens, on roadsides and against farmhouses, up bamboo poles, trellises and fences, and even on the slopes of active volcanoes. The variety of wines is also striking. A lifetime of assiduous study would barely be enough to let one know them all. Such a study, moreover, would be futile, for some wines change from year to year, and many are made by devoted but ill-equipped amateurs, whose chief concern is to have something with which to drown their share of the sixteen million tons of *pasta* that Italians devour each year.

Yet, though many Italian wines are ordinary and undistinguished, there are dozens of honest *vini de pasto* (table wines), full of character despite their modest cost. And the *vini tipici,* which represent the Italian effort to have something like the French wines of *appellations contrôlées,* can be excellent and full of fire—though not even the best of them claim the supreme race and distinction of the great French Burgundies and Clarets.

CHIANTI

One of the world's best known and probably most widely imitated wines, Chianti, is Italian. The classical Chianti region, among the hills, mountains, fields and valleys of Tuscany, is small—some 170,000 acres between Florence and Siena (including the townships of Redda, Caiolo, Castellina, Greve, and part of Castelnuovo Berardenga). Usually unnoticed by the traveler hurrying along the beautiful, meandering *autostrade del sole* ("highway of the sun") between Milan and Naples, it is a wild-looking, rocky region. Here for generations the *Chiantigiani,* or local peasants, have been patiently digging and breaking stone to plant *filari,* single rows of vine, wherever room could be found.

Most Chianti still comes from *filari* or from tiny parcels of land where mixed culture *(coltura promiscua)* predominates. Only in recent years have the large wineries—such as those of Barone Ricasoli, Marchese Antinori, Conte Serristori and other landed gentry—started to shift toward specialization, the *coltura specializzata,* in regular, well-organized vineyards. This is one of the signs of changing times.

The Fattoria Antinori, for example, is one of the oldest wine-producing firms in Italy. Archives in Florence show that in 1385, "Giovanni di Piero Antinori, of the Santo Spirito district," was signed up as an apprentice vintner on the rolls of the Vintners' Guild. The Antinoris were engaged in banking and the silk trade but even then they were producing some wine in the Chianti region, chiefly for their own use and for their guests and friends. As their vineyards slowly grew and their wines became known, the family became active in the wine trade, setting up offices on the ground floor of the Palazzo Antinori, a stately Florentine Renaissance family palace built five hundred years ago by architect Giuliano da Maiano. The Antinoris shipped their wine throughout Italy and also sold it by the jugful at the palazzo, through a small opening made for that purpose in a door on a side street.

Today the *Fattorie dei Marchesi Lodovico e Piero Antinori* export wine the world over. The family tradition is maintained today by Nicolo Antinori, the present Marchese, and his son Piero. The Marchese remembers that only a few decades ago wine making in the Chianti district was pretty much a family affair. Peasants, some of whom owned no land at all, were given a few acres to till, to plant and harvest and, in exchange, shared the yield with the owner. This traditional *mezzadria* (half-and-half split) is now disappearing. Farmers own more land, often unite to form a cooperative, and set up *fattorie* of their own.

Many of the *Chiantigiani* themselves are disappearing. "The young people don't seem to like the country any more," says the Marchese. "Every year, more of them leave for the big city, to take factory jobs or try their luck elsewhere, while the old ones stay home. Now I have on our estate several empty, abandoned farms. A few years ago, more than two million people in Italy were engaged in viticulture. Now, less and less. And of course, our methods are changing with the times. No longer does one plant vines among the olive trees together with beans or cabbage: the new economy calls for larger vineyards with nothing else upon them but vines, to be kept up by a well-mechanized, small work force."

On the Antinori estates of Santa Cristina and Paterno, some 750 acres in the heart of the Chianti *classico* region, many new vines are being set out—sometimes at the expense of century-old olive trees, torn out to make room for the tractors. "It is sad, but we have no choice," says the Marchese.

The "Gallo Nero"

So many imitations sprang up after Chianti became popular that the producers of the Chianti *classico* district joined together in 1924 to form a protective association, the *Consorzio per la Difesa del Vino Tipico del Chianti e della sua Marca di Origine.* Today all the wine produced in the original *classico* region bears the *consorzio's* trademark, the *Gallo Nero,* a black cockerel in a gold medallion surrounded by the association's name.

The red wines belong to either one of two categories which are used to distinguish nearly all Italian red wines: *vini da pasto,* light, fresh, young table wines, ideal to accompany spaghetti or other pasta dishes; and *vini d'arrosto* ("roast wines"), heavier, full-bodied, well rounded and more delicate wines, coming from the best parcels of land. These wines should mature several years in the cask and also in the bottle.

The best-known Chiantis are lighter, lesser wines, nearly always marketed in straw-covered flasks, the *fiaschi.* The Chiantis *per arrosto* come in claret-type bot-

FLASKS BEFORE WICKERING. Here they are stored out of doors. *(Photo Robert R. Weber)*

tles, and are only slowly becoming known to wine lovers outside of Italy. Many wine buyers still suspect that any Chianti not in *fiaschi* is an imitation. In fact, the opposite is truer: the imitations nearly always come in *fiaschi*.

The best of the "classical" Chiantis come from the hills around a dark, imposing medieval castle, Castello di Brolio, for which many a battle was fought between the republics of Florence and Siena. There is a wonderful panorama from the crenelated towers and the 40-foot-high walls which dominate most of the Chianti district. And looking south, one sees the hills sloping down to the Arbia river, and nearby the imposing Mount Amiata and the town of Siena.

Brolio wines, produced by the *Casa Vinicola Barone Ricasoli*, are probably the world's most widely known Chianti wines—either *da pasto* in the flask, or *per arrosto* in the straight bottle. The Castello Brolio Riserva is a truly excellent wine, produced only during favorable years, from the best grapes around the castle. It is full-bodied, austere and harmonious, and is aged at least five years.

According to recently adopted legislation, a Chianti wine labeled "reserve" must be more than three years of age, while a wine said to be "aged" should be at least two years old. This new legislation, as well as earlier regulations adopted at the request of the Chianti growers, also controls the type of grapes that go into the wine. The almost invariable combination for Chianti consists of four varieties of grape: 70 percent of Sangiovese and 15 percent of Canaiolo Nero, both red wine grapes: and two white varieties, Malvasia and Trebbiano, which account for the remaining 15 percent or so. Several Tuscan districts, in Chianti but outside the *classico* region, are entitled to the Chianti label, but not to the exclusive *Gallo Nero*. These neoclassical wines have their own *consorzio* and are bottled under the *Putto Bianco* crest, which represents a cherublike choirboy.

These wines are given the names of hillsides or districts where they are produced: Chianti dei Colli Fiorentini, Chianti dei Colli Senesi, dei Colli Arentini, de Montalbano, di Ruffina, or delle Colline Pisane. Montalbano and Ruffina are usually the best (Ruffina is a small town, but its Chianti should not be confused with Ruffino, a brand produced by the firm of that name).

Some white Chianti is also made from the Trebbiano grape. It is dry, fairly strong, and has a pleasant bouquet. In addition to Chianti, Tuscany produces a few other interesting wines. Among them are two fine reds, Montepulciano and Brunello di Montalcino. Both are "roast wines," and both age well. The Montalcino can reach a splendid ripeness when well aged.

The heart of Tuscany is Florence, center of the Italian Renaissance. It was also in Florence, at the Medici court, that Italian cuisine was born. There is a wide range of specialties "alla Fiorentina": *baccala* (cod) prepared with olive oil, garlic and pepper; assorted *costata*, cutlets fried in oil; the famed *faggioli*, beans, with herbs, garlic and onions; and *funghi*, the fleshy mushrooms, resembling the French *cèpes*, prepared in a variety of manners, usually cooked in olive oil.

PIEDMONT

Piedmont ("foot of the mountain") is a transition area between the rest of the world, beyond the Alps, and Italy—a fitting introduction to the contrasts further south. After driving into Italy from Switzerland or France through several mountain passes and seemingly endless tunnels, the traveler emerges into a country not much different from the one he has just left. This is not yet the land of the *dolce far niente:* The Piedmontese are the busiest Italians, and their city, Turin, is the capital of the automobile and artificial silk industries.

In northern Piedmont, south of Lake Maggiore, there is a wine with a translucent red color and lively orange reflections, light bouquet, deep flavor, and a

ITALIAN WINE LABELS, 19th century. *Mella collection.*

fairly high alcoholic content. It is Gattinara, a delightful wine, most of which is drunk on the spot. It comes from the Nebbiolo grape, a variety used to make several of the better Italian wines. The name comes from *nebbia,* or fog, for Nebbiolo gives best results in districts, such as northern Piedmont, where mornings before harvest time are misty.

But most of the Piedmontese wines are made further south, beyond a range of hills, the Langhe, which stretch into Liguria. This is where Italy's greatest wines, Barolo and Barbaresco, come from. They are made around Alba. Always aged in the wood before bottling, the strong Barolo wines, with up to 15 percent alcohol, are comparable to the great wines of the French Rhone River valley; but they are much scarcer, since only 600,000 to 700,000 gallons are produced every year. Arturo Marescalchi, one of Italy's top wine experts, has aptly described the typical Barolo wine: "In its youth, it is rough and rude, but after three or four years of intelligent refinement it becomes tender, velvety, full of a very agreeable taste and with a bouquet that has, shall we say . . . the violet at the surface, and tar at the bottom."

Barbaresco, also made from the Nebbiolo grape, is lighter, less full-bodied, though not unlike Barolo. It is, says Marescalchi, ". . . feminine and well rounded out: something like the result of a cross between an austere and rigid Piedmontese of the old school, and a nice girl of the hot Emilia region, expansive and graceful." The wine is named after the town of Barbaresco. Another hearty but lesser wine produced nearby is named after a variety of grape: Barbera. It is also full-bodied and has a strong bouquet—similar to, though less delicate than, that of Barolo, and is best drunk in its youth. Grignolino, made in the district of Monferrato ("furious mountain") on the south side of the fertile Po River valley, also borrows its name from a variety of grape. It has a very special flavor and a *frizzante* sparkle that gives it a misleading appearance of innocence—it is actually a fairly strong wine, containing around 14 percent alcohol. There are many other examples, for the steep, rolling, chalky hills of the Monferrato district produce more table wines, and also more sparkling wines, than all of the United States.

The town of Asti, south of Turin on the majestic Po, is best known for its white sparkling wine, incredibly popular in Italy and even exported, of all places,

to France. There are, in fact, not just one Asti Spumante but many, ranging in quality from quite good to sweetish and totally uninteresting. The name of Asti has also come to denote a special type of Muscat wine, made chiefly from the Canelli Muscat, sometimes blended with some Pinot or Riesling to tone down the strong Muscat flavor. If made in the traditional Champagne method, and when not too sweet, it can be quite pleasant, particularly when served very cold, being innocuous, and very light, with from 6 to 9 percent alcohol. The fresh taste of the grapes is one of its great charms, and it should not be aged. It is a favorite among the ladies, and is often served to children at Christmastime, during *panettone e spumante* (light cake and wine) parties.

Vermouth

The Vermouth of Turin is known around the world, either as an aperitif to be taken with ice, perhaps with soda water, or as an almost infinitesimal component of the dry Martini, which should, according to some traditionalists, be made in the city of New York with English gin and Italian Vermouth. Vermouth was first produced in 1786 by a Turin wine maker who tried to recreate, with various herbs, the *absinthianum vinum,* a favorite of ancient Romans. Whether the result is similar or not will forever remain a mystery, but Vermouth (from the German *Wermut,* or wormwood) has acquired a worthy reputation of its own. The original Vermouth is (or should be) made with Muscat grapes.

LIGURIA

Geography has not been fair to Piedmont, which comes close to but never quite reaches the Mediterranean. Liguria, south of it, is more fortunate. It is a crescent-shaped maritime province, where the traveler is seldom more than a few miles away from the Riviera shores. The route goes from Vintimiglia, on the French border, to the port of La Spezia, along a road which has been in constant use since Consul Emilius Lepidus had it built in 187 B.C., to link Arles with Rome.

Notched with deep valleys opening into the sea, Liguria is, of course, another wine-growing province, with many of its terraced vineyards dating back to Roman times and earlier. Ligurians are steeped in all sorts of traditions, inherited from Greeks, from Oriental pirates, and from Romans. Each village has its own costumes and its own feasts. In the Bisagno and Pelcera valleys, for example, the *ciocco* rite is still alive: every Christmas a laurel tree is burned, and its ashes distributed and kept for luck.

Southeast of Genoa, between La Spezia and Levanto, there lies one of the most unusual wine-growing districts in the world, producing a large amount of fair to good white wine. It is Cinque Terre (Five Lands), a group of five small villages wedged between the sea and a mountain. They are best reached by boat, and only recently has a road been built, bringing to the southernmost village, Riomaggiore, the screeching of tires and the roar of tiny, furious engines so familiar elsewhere in Italy.

VINE PROPS IN LIGURIA. *(Photo Luthy)*

The Cinque Terre vineyards are entirely man-made—built some 2,000 years ago by patient Ligurians who carried soil up the rocky mountain, anchored it with stone terraces, and planted vines. Every inch of the rocks seems to be covered with vines, and the thousands of terraces across the mountain give it the appearance of a huge, multilayered cake, green in the spring, golden and red in the autumn. Earth is still carried up the stony mountain paths, and grapes descend in heavy baskets, carried by men on their shoulders or by women on their heads. Some of the terraces are built over ledges so steep that they can be reached only with the help of a rope.

Wine making in Cinque Terre has remained highly individualistic. Nearly each house has its own *cantina* (winery), where one still has the opportunity, rare nowadays, of seeing the wine pressed by foot. Most of the wine, and the best, too, is white, with a bright golden-yellow color. Cinque Terre wine can be bought all over Italy, usually in a bottle covered with a symbolic fishnet, but so much wine is sold under the label that it should be drunk with scepticism.

LAKE WINES

Milan is the capital of minestrone soup, of Gorgonzola cheese, and of ice cream. It is a gourmet town and, though not much wine of any consequence is grown around it, nearly all the best-known Italian wines can be tasted in town.

Interesting wines are made in the valley of the Valtellina, north of Lake Como near the Swiss border. Uninterrupted vineyards clinging to hillsides bespeak centuries of care. Some of the wines, in fact, were praised by Leonardo da Vinci.

WINE CUPS AND GOBLETS, Venice, 16th and 17th centuries. *(Photo Chapman)*

The best, Sassella and Inferno, are red, produced from the Nebbiolo grape, called here the Chiavennasca.

Lake Garda, largest lake in Italy, was given its name by Charlemagne, who made a county of the area. The northern part of the lake is wild and narrow, with torrents falling from the Alps, 6,000 feet above. Further south it becomes wide and forks out, and one of Italy's best rosé wines grows here—the unpretentious, little-known Chiaretto (or Chiarello) di Garda. Light, fresh, prettily pink, it is undoubtedly best when drunk overlooking the lake, in hotels or restaurants or village *trattorias,* where it is served not in the bottle but as *vino aperto,* "open wine," in a rotund decanter.

Veneto, the region of Venice, produces large quantities of wine. There are a few outstanding growths, notably from the area of Verona—the town of Romeo and Juliet. Valpolicella is a red wine, the best of which is aged in wood for at least a year before bottling. Light and fruity, the Valpolicella Superiore can then improve in the bottle for up to five years. But there are many lesser growths, served as *vino aperto,* that can be delightful. Four grape varieties are used: Corvina, Negrara, Molinara and Rondinella. Made nearby, chiefly with the same grape varieties, Bardolino is much lighter in color, and best when very young.

Another well-known wine made near Verona is named after the small town of Soave. It is one of Italy's best white wines. Pale, light, occasionally lacking in acidity, it should be drunk when young. From Alpine valleys to the north come small quantities of delicious wines—Riesling, Pinot Grigio, Merlot—which somehow seem more Austrian than Italian.

SAN MARINO AND THE MARCHES

Italy is rich in historic memories. On the way from Milan to Rimini, the traveler may pass through the historic old town of Canossa and, near the Adriatic coast, will cross the tiny Rubicon, as Caesar did in 49 B.C. And there, near the end of the route, is San Marino, one of the world's smallest (38 square miles) but oldest republics. It was founded in the 4th century by a mason called Marino, and the first extant document testifying to the independence of the republic dates back to 885. Perched atop a rocky spur, the tiny state imports much wine from Italy, but manages to produce some of its own—notably, the delicious, fruity white Moscato of San Marino.

Still further south, the region of the Marches stretches along the Adriatic. This is the home of the superior white wine, the Verdicchio. It is said that when Hannibal crossed the region with his soldiers, Verdicchio threatened to destroy his strength. Yet it is one of the lightest white wines in Italy, particularly pleasant on a summer day. Sold in a characteristic green, club-shaped bottle with a golden cap, Verdicchio has gained much popularity in recent years, and some have claimed it is marketed in amounts exceeding the yield of the available Verdicchio grape.

AN ETRUSCAN VASE in the shape of a leg. Bucchero pottery in relief, 6th century B.C. *Louvre. (Photo Chapman)*

Left: PICKING GRAPES. Vines are growing on tall props, a practice common in Italy. *(Photo Mondadori-Canieri)*

ETRUSCAN WINES

The Etruscans knew how to take care of their vineyards. The Roman agronomist Columella has it that a single plant could bear as many as a thousand bunches, and Pliny the Elder reports seeing, in Populonia, the former Etruscan capital, a man-sized statue of Jupiter hewn out of a single vine stalk. In order to keep thunder, disease, hailstorms and bad luck in general away from their vineyards, the Etruscans used to protect them with the skull of a donkey or sheep —a custom still not extinct.

Driving along the road which cuts across the Italian boot toward Rome, the traveler will reach Orvieto, once an Etruscan city, which calls to mind the image of a cool, rotund, and squat *fiasco* of white wine, waiting to soothe the parched palate. This district is the site of extreme *coltura promiscua*—so promiscuous in fact, that the vine has to be looked for, growing around trees, among string beans, in cabbage patches, against walls, or hanging from fences. Light on the palate and pale in color, Orvieto wine has a subtle bouquet, and is best drunk when young. The *secco* is dry, pleasing, and thirst-quenching, while the *abboccato* loses some of its freshness to its sweetness.

Still further south, toward Rome, there is another fruity white wine produced near the aptly named town of Montefiascone ("mountain of the big flask"). Called "Est! Est! Est!", this wine has been famous since 1111 when a German cardinal, Johannes de Fugger, traveled to the Vatican. To insure that he would find good wine to drink along the way, the cardinal dispatched before him a valet, well trained as a taster, and instructed him to write on the walls of taverns where wine was particularly good the Latin word "Est", meaning "It is." Reaching Montefiascone, the valet wrote "Est! Est! Est!", a hyperbolic emphasis meant to indicate that the wine was outstanding. The valet continued on his way to Rome, but the cardinal never made it—when he reached Montefiascone, he drank so much wine that he died. The disbelieving can visit his well-preserved tomb in the San Flaviano Church, where the story is mentioned in an epitaph in Latin.

FROM ROME TO NAPLES

Unlike the cardinal, the wine lover may now reach Rome, and its excellent and romantic restaurants, boasting a varied assortment of wines to accompany the favorite Roman dishes: *Saltimbocca,* thin slices of veal cooked in butter, surrounded with ham, doused with Marsala and perfumed with sage, whose name aptly means that they "jump into the mouth"; *abbacchio al forno,* lamb cooked in the oven; or *carciofi alla giudia,* small artichokes cooked in oil with garlic and parsley; and many others.

A brief excursion out of Rome is in order to visit the *Castelli Romani*—the Roman Castles—and their wines. The traditional playground of wealthy Romans ever since Rome's origin, the *Castelli* owe their name to thirteen castles erected only in the Middle Ages. The most celebrated and elegant is Frascati which produces the best wines. Cicero wrote, Lucullus ate, and Tiberius drank in Frascati, today the home of Cinecitta—the Italian Hollywood—and of an unpretentious, dry, full-bodied, rather strong wine.

From Rome, Naples can now be reached via the *autostrada*—or, better yet, by sea in order to discover the beauty of the seashore, the islands, the tiny villages, and Mount Vesuvius puffing smoke. The city of *bel canto* is of Greek origin, and Neapolitans are small, dark, with a Hellenistic profile and a sing-song dialect of their own, mysterious even to other Italians. Naples is the birthplace of spaghetti—which should be eaten *al dente,* slightly hard, and never overcooked; and of pizza, best tasted in the old city in the evening around a glowing oven.

Falerno wine, so celebrated during the days of the Caesars, and of which Horace claimed to have drunk a century-old vintage, is produced along the coast of Naples. Both red and white, Falerno wines are pleasant but are far from being the divine nectar they were perhaps two thousand years ago. Much better is a white wine made near Naples from Greco della Torre grapes grown on the slopes of Mount Vesuvius. It is called Lachryma Christi, the "tears of Christ." The story goes that when Lucifer was cast out of Paradise, he stole a clod of the celestial earth and dropped it into the Gulf of Naples. This island became known as Capri, a sinful land but a heavenly spot, too. And when the Lord returned to earth he was saddened by the sight of the wayward island, and cried. Where his tears fell there grew vines from which this wine has been made.

A fair red wine is grown a few miles south of Vesuvius, near Pompeii: Gragnano—light and fruity, a bit sweet though not unpleasantly so, and much favored by Neapolitans as wine to go with pizza.

Capri, a two-pronged rocky playground island at the edge of the Gulf of Na-

HARVEST AT CINQUE TERRE. Women bring the harvest down from the steeply terraced slopes in wicker baskets carried on their heads. *(Photo Dorozynski)*

ples, whose golden rule is to "Do as you please," is said to have one private villa where a small fountain, turned on during informal gatherings, fills a pool with sparkling wine. The wine of Capri, dry and white, can be excellent, but much of it doesn't come from Capri itself—the producing district legally includes Capri, the island of Ischia and a stretch of the mainland. The quality of the wines varies greatly (the best is from Ischia), depending upon where it comes from—but the source is seldom indicated on the bottle.

THE HEEL, SPUR, AND TOE . . .

Once more crossing the land, the traveler now reaches Apulia, the "heel" of the Italian boot. Rugged but not without beauty, this little-known region has something hidden and secret about it. The land is hot, dry, and arid, suitable chiefly for the growing of almonds, figs, olives and grapes. Wine growing was started here by the Greeks some seven or eight centuries before our era, and Apulian wines have been exported since the 4th century B.C. Today, this is the "wine cellar of Italy," planted chiefly with undistinguished assortments of high-yield grapes. This is a mass production vineyard, producing between 100 and 150 million gallons of wine a year, most of it wines for blending which emigrate in tankers or in trucks to foreign lands. The huge cooperative wineries make a striking contrast with the *trulli,* the typical, beehive-style one- or two-room stone houses inhabited by laborers and farmers—who spend, it seems, most of their their life sleepily traveling from vineyard to cellar to village on the backs of un-

A WINE MERCHANT. Lithograph, early 19th century. *(Photo Etienne Hubert)*

hurried donkeys. A few Apulian wines, however, have some finesse in their rustic heartiness. The dry, pale-straw-colored Sansevero is light and fresh and ages well for a few years. Lacorotondo, once used exclusively to make Vermouth or to be blended with other wines, can become quite palatable and stand on its own if it is given time to age, and so can the white Verdea wine from Alberobello.

A few dessert wines made with Aleatico, Moscato and Malvasia Blanco are pleasant, notably the Moscato di Trani, which contains 16 percent alcohol and 14 to 15 percent sugar. And around the village of Cerignola, just under the "spur" of the boot, a small inland wine-growing district produces an excellent, potent, slightly harsh wine, the Torre Quarto.

Not far from Apulia is Calabria, the "toe" of the boot. This is Italy's poorest, most desolate region, with few economic prospects except for its developing Riviera. Calabria has a wild beauty which appeals to lonely souls who like to view the world below from high cliffs rising steeply above the sea. The vineyards stretch from Reggio to Tropea, and the wines tend to be utterly uninteresting, with a few minor exceptions. Greco di Gerace is a golden-amber wine with the perfume of orange blossoms, and a potent alcoholic content—up to 17 percent. Moscato di Cosenza, another dessert wine, is not unlike the Greco.

THE ISLANDS

Sicily is one of the world's hottest wine-growing regions, and its wines are strong, usually sweet, usually very aromatic. Most famous is the Marsala, made only from specified grape varieties (Catarratto, Grillo, Inzolia) grown in a limited region around the seaport of Marsala. Marsala is the former Lilybaeum, the principal fortress of the Carthaginians in Sicily. It was destroyed in 1532 by King Charles V of the Holy Roman Empire, because he feared he would not be able to defend it against the Turks. The Turks came, gutted the town, and named it *Marsah Allah,* "God's Harbor." The wine industry in Marsala is still largely in the hands of British firms. It was started by one John Woodhouse, wine merchant of Liverpool, who came to Sicily in the 18th century to create an inexpensive substitute for Spanish and Portuguese wines, then in great demand in England. It was Woodhouse who supplied wines for Nelson's fleet in 1800.

There are several types of Marsala, the best and driest being Vergini, aged and blended in *soleras* like Spanish Sherry (see SPAIN, *Solera*). A few fair table wines are also made in Sicily, notably those produced from vineyards 2,000 feet up on the slopes of Mount Etna, the active, unpredictable volcano whose crater nowadays can be reached by cable car. Two rather good wines are grown on the volcano's lower slopes, and in the neighboring village of Catina. Etna Bianco is yellow-greenish in color with golden reflections and a rich bouquet. Etna Ross is vividly bright, dry and well balanced, and is aged for at least three years in barrels of Slavonic oak.

North of Sicily are the Lipari Islands, also called the Aeolian Islands, because the ancients believed that here was the home of Aeolus, the god of the winds.

WINE LABELS of Sicily and the Aeolian Islands. *Mella collection.*

The largest of the group, Lipari, has a celebrated wine, Malvasia. It is very sweet, heavy, amber-colored and long-lived, coming from a much-traveled variety of grape, originally named Monemvasia from the Greek harbor of that name. Some of the wine comes from grapes grown on the edges of Stromboli, a smallish, well-behaved volanco-island which calls attention to itself every other hour or so by blowing up a few stones and some lava. Of the seven Lipari islands, Salina produces the most wine, which also goes under the name Lipari.

In Sardinia the most noteworthy wine is Vernaccia, an amber-colored natural liqueur which defies comparison with anything else, and which you frankly like—or don't. Vernaccia is one of the world's strongest natural wines, reaching an alcoholic content of 17 or 18 percent, and it has a very particular bittersweet taste. It can be drunk as an aperitif, but some like it as a table wine too. Unfortunately, much Vernaccia wine on the market is not Sardinian, for its fame has encouraged imitators. Vernaccia was mentioned by Chaucer in the Merchant's Tale, as a wine "Of spyces hot, t'encressen his courage," and by Dante, who consigned Pope Martin IV to purgatory because he stewed Bolsena eels in Vernaccia wine.

In Italy wine is so much a part of daily life that one doesn't make much fuss about it. Order a white wine with your steak, and it is likely that the waiter will produce it with a smile and without the bat of an eyelash—while in France such a request will bring about at least a noticeable frown. As the Marchese Antinori puts it: "Perhaps we don't make the best wine in the world, as the French do, because in France wine making is looked upon almost religiously. In Italy, it is fun—and essentially human."

A CANTINA PINAROLI LABEL.
Mella collection.

ZELL, in the Moselle vineyards. *(Photo Bildarchiv Irmer)*

GERMANY

There are a number of good reasons why Germany should not produce any good wines. German vineyards are the world's northernmost, straddling the 50th parallel approximately on the latitude of Newfoundland, Labrador, Winnipeg and Vancouver Island. The vineyards are frequently threatened by frost, and many of the best grow along precipitously steep slopes. Moreover, Germans are not, by and large, wine drinkers. They down steinsful of excellent beer in great quantity, while the average personal wine consumption is between two and three gallons a year—a reasonable weekly allowance for a French or Italian laborer.

Yet, Germany manages to export every year some three to four million gallons of wine, among which are some of the world's finest, most distinctive white wines—the "Hocks" and Moselles, for which the demand is greater than for any other type of white wine.

Most of Germany's wine-growing land was part of the Roman empire for many centuries, following Julius Caesar's conquest and the founding of *Colonia Augusta Trevirorum* (modern Trier) by the Emperor Augustus.

Flourishing in the first few centuries of the Christian era, the German wine industry declined from the time of the Roman withdrawal until Charlemagne, who gave it a new lease on life. From his palace of Ingelheim on the southern bank of the Rhine, the emperor had observed that on the opposite bank, facing south and protected by the Taunus mountain range, snow melted much faster

BARREL HEADS, 18th and 19th centuries.
Speier Museum, Germany. (Photo Chapman)

than elsehwere. He therefore ordered the planting of vineyards in the district known today as Johannisberg.

Since Charlemagne, many vineyards have been owned by the church, which long held a virtual monopoly on the best German wines. The archbishop of Mainz owned nearly all of the Rheingau; Benedictine monks owned Johannisberg, and Cistercians laid out the vineyards at Eberbach. On the Moselle, Benedictines of St. Maximin Abbey laid out many vineyards around Trier, while St. Eucharias Abbey, of the same religious order, owned excellent vineyards down the river, in Bernkastel, Kues, and Trittenheim.

German wines enjoyed great success in England, where they were indiscriminately referred to as "Rhenish," even if they came from the Moselle or from Hesse. Hochheim on the Rhine was an important shipping center, and Bacharach (named after Bacchus) could be reached by the larger wine ships. Queen Victoria much appreciated Hochheim wine and since her time "Rhenish" has been replaced in England by "Hock," a contraction of Hochheim.

In 1874, the first phylloxera damage was reported in the Palatinate, and many of the vineyards, destroyed by this insect pest in the decades that followed, were not replanted. Much fraud was perpetrated in attempts to supply the world with Hock but, starting at the beginning of the century, new legislation has restored order by regulating production, treatment, and labeling of wines. Since World War II, many fragmented vineyards have been put together and organized to permit modern machinery to work the vines. And thanks to the planting of new vineyards, the output of wine in Germany has nearly doubled in recent years.

Most of the finest German vineyards, however, have remained small, and German winegrowers tend to be as fussy about their wines as Swiss watchmakers are about their timepieces. A single hillock, which may yield a few thousand gallons of wine, is meticulously subdivided so that the wine is often bottled under three or four different labels, each representing a parcel of land with a slightly different exposure to the sun, a slightly different slope or soil. As Germany has some 10,000 place names for wines and a dozen special ways of making wine, the number of different labels can exceed 30,000!

In selecting German wines, it is better to look for those that have not been subjected to chaptalization, or special sugaring to compensate for lack of natural sugar in a poor year. Unsugared wines are usually finer and more subtle. These wines are identified by the words *Naturwein Originalabfüllung*, or *Natur* or *Naturwein*. In many cases, when winegrowers belong to the Natural Wine Association, their membership is indicated on the label. This is done by means of a seal

THE GIANT KURFURSTEN HOFF WINE KEG, HEIDELBERG,
on view between 1589 and 1591. Engraving, 17th-century. *(Photo Arborio Mella)*

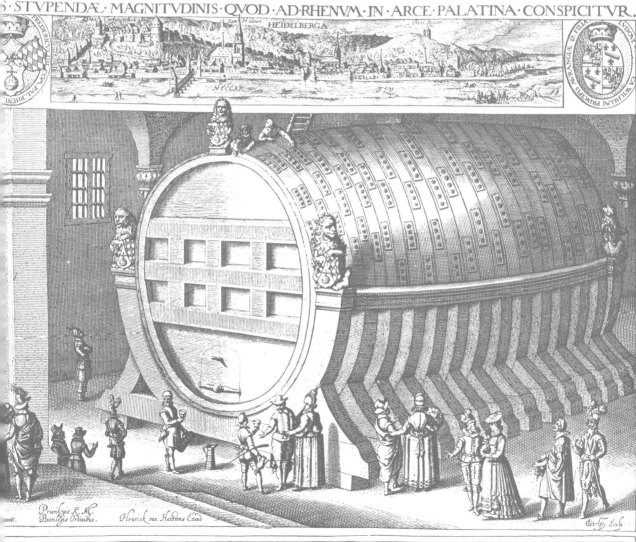

WATCHMAN in the Rhineland vineyards.
(Photo Chapman)

1,200 YEARS OF WINE GROWING—
sign at a vineyard in Oppenheim.
(Photo Chapman)

representing an eagle, whose body is made with a bunch of grapes, and by the name of the association, *Verband Deutscher Naturwein-Versteigerer E.V.*

The label nearly always gives the name of the village, town or vineyard where the wine was produced, as well as the name of the region. Better wines have the name of the vineyard, the owner or the *Schloss,* and the indication that the wine was bottled by the owner. *Gewächs* or *Wachstum* means growth, *Weingut* means winery, and the fact that the wine was estate-bottled is indicated by the words *Originalabzug, Originalabfüllung,* or, in the case of a castle, *Schloss-Abzug.*

The label often indicates the grape variety used for the wine, especially for Moselle and Rheingau wines, where the Riesling grape is used almost exclusively. A complicated system of colored bottle caps is frequently used to differentiate between the qualities of different wines from the same domain, but unfortunately they don't seem to follow any fixed rule. The word *Kabinett* is traditionally used for wines the producer considers to be his best. However, the term has no legal significance. Neither have such words as *feine, feinste,* and *hochfeinste.*

"SPECIAL" WINES

The label may also indicate that one of many special processes has been used to make the wine. German growers go to extremes to make the best of their harvest, and are particularly proud to produce one, or several, "special" wines.

Spätlese (late harvest) means that the grapes from which the wine was made were left to ripen after the regular harvest had been completed. The grower takes a chance with the weather, hoping to increase the sugar content of the grape in order to have a more full-bodied, stronger, sometimes sweeter wine.

Auslese means the wine is made only from the ripest grapes. In a good year, this means a sweeter, heavier wine, with a more concentrated bouquet. In a poorer year, it indicates the wine may have less acidity than the rest of the har-

DAWN OVER THE RHINELAND at harvest time. *(Photo Chapman)*

vest—or that *Auslese* only is natural, while the rest of the harvest has been sugared.

Beerenauslese is rare, liquorous, and expensive. It is made from the selection of single grapes or packets of grapes snipped off with scissors and pressed separately, usually under the jealous and loving eye of the owner, to yield an essence of sweetness, bouquet, fruit, and flavor comparable to that of a great Sauternes.

Trockenbeerenauslese is the proudest achievement of the German winegrower. In exceptional years, perhaps once every decade, *Botrytis cinerea*, the same mold that withers Sauternes grapes, covers some of the grapes with tiny holes, through which water evaporates. Chosen bunches are laid out on a table and from them, the driest (*trocken*, dry), most raisinlike grapes are picked out by hand. They yield a small amount of heavy, amber-colored must, which—once it has fermented into wine—enthusiasts claim to be worth its weight in gold. A very sweet, lavish and liquorous wine, yet not cloying, it is not comparable to any other growth.

Since World War II, only a few years—including 1953, 1959 and 1964—have been sufficiently propitious to permit the making of good *Trockenbeerenauslese*.

Edelbeerenauslese is a wine made from grapes which have been shriveled by *Botrytis cinerea* and picked out as for *Trockenbeerenauslese,* but which are not quite as dried up. A fine *Edelbeerenauslese* is made in the Ahr valley (probably the world's northernmost commercial vineyard) from red Burgunder grapes.

Yet another highly prized German specialty is *Eiswein*. The tradition of making this "ice wine" dates back to the chill autumn of 1842, when growers in Traben-Trarbach on the Moselle were grieved to see their grapes frozen hard in the vineyards by an untimely frost. They harvested the grapes nevertheless, but pressed them separately, for fear of spoiling the rest of the vintage. To their surprise, the resulting must, though sparse, was exquisitely sweet and pure, and the wine comparable to the *Auslese* of a good year. What happened, in fact, was that the frost, which must have reached only a few degrees below the freezing point,

GERMAN WINE LABEL
showing the town of Hochheimer,
19th century. *Mella collection.*

STREET SIGN in Rhöndorf am Rhein. *(Photo P. Strack)*

had turned the water in the grape into ice crystals, at the same time increasing the sugar and flavor concentration in the unfrozen pulp. Since then, many growers have occasionally attempted to produce ice wine, though there is always a risk involved: the grapes can rot before they freeze. And if the temperature falls too low, the harvest can be lost.

St. Nikolaus, Germany's Jack Frost, is honored on December 6, and if the weather appears to be favorable, the growers delay the harvest until that day, to make a special *Spätlese* called St. Nikolaus Wein. Much rarer *Spätlese* wines are the Sylvester wine, gathered on New Year's Eve and *Dreikönigswein,* "Three Kings' Wine," gathered at enormous risk on January 6, or Twelfth Night.

A popular type of wine—never bottled, but served straight from the cask—is *Federweisser,* new wine whose fermentation is not yet completed, and whose yeast and active carbon dioxide give it a milky appearance. It has the reputation of helping digestion—and going to the drinker's head.

Maiwein, traditionally drunk in May and June, preferably in Bavaria, is spiced with woodruff, and is served cold usually with strawberries floating in it. It has a special flavor of its own which one does or does not like. *Jungferwein* ("virgin wine") may owe its existence to the destruction of many vineyards by the phylloxera. The Germans maintain that a vineyard upon which no vine has been grown for several years has had the chance to accumulate exceptionally rich nourishment. Then, three or four years after new vines are laid out, the first harvest is said to be "virginal." The wine, usually drunk in its youth, can have a particularly fresh, pungent bouquet and taste.

Finally, there is the German sparkling wine, known as *Sekt,* the origin of the name being either a distortion of the French *sec* or a misunderstanding of the

A HOCHHEIMER WINE LABEL,
19th Century. *Mella collection.*

SIGN IN THE RHINELAND
announcing the arrival of new wine. *(Photo Chapman)*

English "Sack." *Sekt* can be made by the traditional "bottle-fermentation" method, by a bulk process, or by simply pumping gas under pressure into the bottled wine. Sekt becomes particularly abundant in poor vintage years, when many producers decline to bottle their wine because it is not worthy of their name, and sell it to Sekt factories. Sekt can, however, occasionally be quite good.

KING RIESLING AND COHORTS

The famed Riesling plant, which dates back at least to Roman times, is the almost exclusive ingredient of the greatest German white wines. Riesling is also one of the world's most widespread plants, covering hundreds of thousands of acres in Europe, South America, the United States and the Soviet Union. Unfortunately, too, it is one of the most misused names in viticulture, given to at least a dozen varieties of plants, some of which—such as the Missouri Riesling—have not the remotest relationship to true Riesling.

One of the most abundant white wine grape varieties in Germany is the Sylvaner, which yields more wine than the Riesling but none as long-lasting or distinguished. The Müller-Thurgau, which is a cross between Riesling and Sylvaner, also gives quite ordinary wine, as does the Elbling. Traminer and Gewürztraminer give rather light wines with a pronounced aroma, the latter being the spiciest. Gutedel is the same as the French Chasselas and the Swiss Fendant. In the south, it yields a pleasant, light, innocuous white wine. The Burgunder (Pinot) and Portugieser are the predominant red wine varieties.

The relatively cold climate in Germany does not give the grapes a high sugar content; hence most of the wines are low in alcohol, seldom exceeding 11 percent. At best, this gives them freshness and contributes to the bouquet, and at worst it makes them too tart, too acid, and lacking in body and alcohol. On the whole, German wines are scrupulously well made. But many nevertheless receive the "sunshine of Frankenthal"—Frankenthal being a large beet-sugar producing center.

The lightness of German wines makes a pleasant contrast with the food, which is rather heavy and usually abundant. The original hamburger *(Deutsches Beefsteak)* and hot dog *(Frankfurter Würstchen)* are mere appetizers in comparison to sauerkraut, beef heart or liver, *Zwischenrippenstück mit ei* (ribsteak and egg), *Klops* (meatballs) and *Spaetzle* (flour-balls), both of which reach the size and weight of a golf ball, and the *Baumkuchen* cake, which can be six feet tall and more.

Rheinland ist Weinland, goes the saying, and indeed the best wines grow along the majestic river, whose banks are dotted with ancient inns. Here amid that certain atmosphere called *Gemütlichkeit,* one can enjoy the local wine, usually served in a large, greenish halfpint *Römer.*

One doesn't usually clink glasses, but austerely meets the eyes of one's fellow drinkers, bows slightly while raising one's glass, and utters, with firm emphasis but without undue enthusiasm, the popular *Prosit* ("Good health!") or the more decorous *Zum Wohle* ("Well-being!"). It takes a few glasses for the dignified drinkers to warm up to the traditional wine songs—*Trink, trink, Brüderlein trink,* or *Ein Rheinisches Mädchen beim rheinischen Wein.*

RHEINGAU

The most complete, well-balanced and distinguished wines come from the Rheingau, twenty miles of slopes along the right bank of the Rhine from Wiesbaden to Rüdesheim, known as the German Riviera. The Rhine here turns to flow west-southwest so that the vineyards on the right bank face almost directly south. Thus they receive the protection of the wooded Taunus mountains to the north, and are warmed by the sunshine accumulated or reflected from the river.

The largest and one of the most famous vineyards in the Rheingau is Schloss Vollrads, where wine has been made by the same family since the 14th century. The head of the family today is Count Matuschka-Greiffenclau, owner of some eighty acres of vineyards that surround the castle on all sides.

All the wines that bear the Greiffenclau name and coat of arms are worthy of attention, and the best are great ones, impregnated with a subtle, delicate fruit which seems to remain on the palate long after the wine itself has gone. A former President of the German Wingrowers' Association, Count Matuschka-Greiffenclau produces on a good year as many as a dozen different wines, sold under the Schloss Vollrads label, and on a poor year (like 1956) none at all. If the wine is not deemed worthy, it is not even bottled in the castle but is sold in bulk to manufacturers of *Sekt,* to become lost in blended anonymity.

The wines that pass his severe test are graded along a range of fine distinctions, sometimes detectable only to experts. The least outstanding wear a plain

PFALZ, ON THE BANKS OF THE RHINE. *(Photo Gandelier-Lacros Atlas Photo)*

green capsule and a Schloss Vollrads label with the words *"Originalabfüllung"*—estate-bottled. Wines of a higher grade have a green capsule with a silver stripe around it; then, a gold stripe. At the next step upward the capsule becomes red, then red with a silver stripe, and red with gold. As they go up, the word *Originalabfüllung* is changed to *Schloss-Abzug:* the castle itself takes the responsibility for bottling. The bottles now wear a green capsule again, then a red one, with finer shades of quality indicated, as before, by a silver and gold stripe; then comes a plain blue capsule, a blue with a silver stripe, and blue with gold. The blue-capped bottles, in addition, bear the word *Kabinett.*

When the harvest is exceptionally good, once or perhaps twice every ten years, a few thousand bottles may receive the white capsule: they are the *Beerenauslese* and, even rarer, the *Trockenbeerenauslese,* truly Olympian liqueurs, and not accessible to many mortals.

West of Vollrads lies Germany's most famous vineyard, around the Schloss Johannisberg, completely rebuilt since World War II. These are the steep slopes Charlemagne observed from his castle across the river, and where he planted vineyards. Once known as St. John's Mountain, the hill was the site of a Benedictine abbey in the 12th century, became the property of Prince William of Orange when the church was dispossessed of its worldly goods in 1801, and was given by Napoleon to one of his marshals, Kellermann. In 1815 the Congress of Vienna gave it to the emperor of Austria, who passed it on to Prince Metternich.

Johannisberg wines are marketed under two different types of labels, one with the Metternich coat of arms, the other with a picture of the castle. The first come under three different capsules: Red for the least outstanding wine, followed

A WINE CELLAR in Rudesheim's famous Drosselgasse. *(Photo Chapman)*

by green (usually a *Spätlese*), and pink, an *Auslese* produced only in favorable years. It is the finest and most distinguished of all Johannisberger wines, usually even better than the wines sold as *Kabinett* with the label of the castle. Only those wines considered to be worthy of the princely owners are bottled at the Schloss—and the rest goes to *Sekt*.

Another celebrated vineyard, and also one of the largest ones in the Rheingau, is the Steinberg, laid out and walled by Cistercian monks in the 12th century, and now owned by the state. Steinberger wines are generally fruitier and more full-bodied than their neighbors, but may be less distinguished too. In good years, they can be superb. The most full-bodied wines come from Hallgarten.

Rüdesheim, a tourist resort peppered with *Weinstuben* and souvenir shops, and the center of 600 acres of vineyards, should not be confused with the small wine town of the same name on the Nahe river, whose wines are of little distinction. Rheingau Rüdesheim wines are generally excellent and varied, the best coming from the Berg, a terraced vineyard above the town, whose wines are labeled as Rüdesheimer Berg. Schloss Reinhartshausen, owned by heirs of the kaiser and now a hotel, has its bottles labeled *Originalabfüllung Prinz Heinrich Friedrich von Preussen*. They are graded by colors which do not correspond either to the Schloss Johannisberg nor to the Schloss Vollrads color scheme. The least distinguished have a label with a crimson border, the better ones, a blue border, and the superior *Kabinett* wines, a red border. Schloss Eltz also ranks among the best, as do the districts of Kiedrich, Rauenthal, Walluf, Hattenheim, Oestrich, Eltville, Winkel, and Geisenheim.

The village of Assmannshausen, north of Rüdesheim, is also part of the Rheingau, and is famed for its red wine made from the Pinot Noir grape, known here as *Spätburgunder*. Full-bodied, flavorful, somewhat astringent, a good Assmannshausen red wine ages well and, in a vintage year, may be comparable to some of the lesser wines of southern Burgundy.

Hochheim, east of Wiesbaden, and actually on the river Main, is the name of another important wine-producing district and town which is legally part of the Rheingau, and whose wines, in the traditional, slim brown Rheingau bottle, have that unmistakable Rheingau character.

FROM THE MIDDLE RHINE TO THE RHEINPFALZ

Downriver from the Rheingau, from Bacharach to Coblenz, the grapes grow on both sides of the river, which has resumed its north-northwesterly course. This is the Middle Rhine, the Rhine of the great castles, once strongholds of robber barons who claimed their tribute from the passing ships; the Rhine of the Lorelei rock, atop of which there sat the flaxen-haired young maiden of German legend who, by her beauty and her songs, lured sailors to their doom. Here grow the vines of Bacharach, Boppard, Oberwesel and Braubach, little-known wines produced in small amounts with great care, with deeper color but less bouquet than the Rheingaus, and well known to those who follow the *Weinstuben* route from Bingen to Cologne.

From the Rheingau south to Worms, dotting the left bank of the rhine, are some 150 wine-growing villages of the Rheinhessen, or Rhenish Hesse. Two categories of wine are distinguishable here: the fine Rieslings on the hills of Bingen, the steep slopes of Oppenheim and Nierstein, and the plain of Worms; and the ordinary, undistinguished table wines, often labeled as *Liebfraumilch*.

The best Hessian wines come from the Nierstein district, and the secret of finding a true Niersteiner is to look on the label not only for Nierstein but also for the name of the vineyard,—for example, Niersteiner Glöck, Niersteiner Hipping, Kehr, Orbel, Gutes Domtal, etc. True Niersteiner Rieslings resemble the great Rheingau wines, although they are smoother and warmer, though less distinquished and less subtle. Directly south of Nierstein is its chief rival, Oppenheim, some of whose vineyards (Herrenberg, Kröttenbrunner, Kreuz and Goldberg) the Riesling shares with the Sylvaner and the Müller-Thurgau. Oppenheimers are excellent wines, sometimes approaching greatness, though never as closely as their neighbor to the north.

Hesse then gently slopes down to the historic city of Worms, where stands the Romanesque cathedral, the Liebfraukirche that has unwillingly given its much maligned name to millions of gallons of often undistinguished *Liebfraumilch*, the "milk of Our Lady." Originally the term was *Liebfraumünch* or *mönch*, indicating that the wine was made by the monks. By legal definition, any Hessian wine of amiable nature and reasonable quality is entitled to be called *Liebfraumilch*. Since the name has gained tremendous popularity, much poor wine is sold under the

STUDENTS DRINKING WINE. Print, 19th century, Germany. *(Photo S. Hano)*

TASTING THE NEW WINE. *(Photo Chapman)*

CUSTOM SAYS, "Sniff it, drink it, then talk about it." *(Photo Kurt Rohrig)*

ANOTHER WINE CELLAR in Rudesheim's Drosselgasse. *(Photo Chapman)*

label- but some good, fresh and fruity wine too, which makes selection difficult. The only procedure is to discover, by trial and error, which merchants have good *Liebfraumilch,* and to remain faithful to these.

The vineyards surrounding the church produce a wine sold as *Liebfrauenkirche* or *Liebfrauenstift;* it is quite fair, though far from the excellence of Niersteiners and Oppenheimers. Other fine Hessian wines are made in Bingen, Dienheim, Bodenheim, Laubenheim, Nackenheim, and Alsheim.

South of Hesse lies Germany's most productive wine district, the Rheinpfalz, or Palatinate, named for the Roman Palatine Hill. The Pfalz was the wine cellar of the Holy Roman Empire, and is still the wine cellar of Germany today. The district lies between a chain of hills called the Haardt on the west, and the Rhine River on the east. Protected by the Haardt, having a mild climate and fertile soil, the Palatinate vineyards have a good yield, producing much ordinary white and red wine. The finest wines come from the Middle Haardt, between Bad Durkheim and Neustadt, in the districts of Wachenheim, Deidesheim, Ruppertsberg, Forst and Bad Durkheim, which is the site of one of the largest festivals in Germany, the Bad Durkheim *Wurstmarkt.* Every year, this traditional fair, dating back to the 15th century, attracts some 200,000 people who drink 50,000 gallons of wine and eat about 100 miles of sausages.

The southernmost German vineyards are in the Baden district, growing along the Black Forest and looking south to the Rhine Valley and the Swiss border. There is a great variety of wines here, many of them pleasant and unpretentious, such as the *Seewein* (lake wine) produced near the Boden See (Lake Constance) and the *Weissherbst* (white harvest), which sometimes has a slight rosé tinge.

THE MOSELLE

The Moselle River has its source in France, follows the border of the Grand Duchy of Luxembourg, and enters Germany near Trier. Many of the Moselle wines are wholly as good as the best of the Rhineland, though entirely different in character. Moselle wines are also produced in France and in Luxembourg, but these never approach the quality of wines grown along the river once it has passed through Trier and flows north to the Rhine at Coblenz.

The story goes that in olden times, night watchmen in the small villages of the Moselle valleys came by every two hours during the night to knock at the doors of the most assiduous drinkers, who then turned over in their beds so as not to expose the same side of their stomachs to wine for more than two hours in a row. Actually, Moselle wines are not all that acid, except in particularly poor years, when even sugar cannot hide their sting. But in good years, they can be splendid, and in exceptional vintage years can even surpass those of the Rheingau, while keeping their own character and personality.

Moselles are pale yellow, sometimes with a tinge of green; they are light, and low in alcohol, delicate yet mouth-filling, fragrant and full of the Riesling bouquet. Fine Moselles (all come in a slim green bottle) make up about a third of the district's annual yield of twenty-five million gallons. They present a complex picture to the prospective purchaser, for here, even more than along the Rhine, the vintners have a maniacal concern for keeping one cask apart from another which might come from a patch ten feet away but is deemed to be ever so slightly different. As a rule, the wines carry the name of a village, as well as that of a specific vineyard. Sometimes, the name includes that of a portion of a vineyard which the owner believes to be better than another part. There are, of course, *Spätlese, Auslese,* and *Eiswein,* as well as *Beerenauslese* and *Trockenbeerenauslese,* which again are made only in exceptional years.

The best of Moselle wines come from the pretty, Hänsel-and-Gretel little town of Bernkastel and, more precisely, from the southern slope of a large hill overlooking it, the Doktorberg (which takes its name from having once been owned by a doctor, and not, as some have it, because the wine has medicinal properties). There is nothing quite comparable to the Bernkasteler Doktor (or Doktor-und-Graben) *Spätlese* of a good year—for example, 1951 or 1963 (the wines should be left to age at least four or five years). Other excellent wines come from the districts of Piesport (with its *Goldtröpfchen*), Graach, Zeltingen, and Wehlen.

The Moselle and its two confluents, the Saar and the Ruwer, have a single

WORMS CATHEDRAL, with surrounding vineyards *(Photo Norbert Seilheimer)*

regional name, Mosel-Saar-Ruwer—a fact that is deplored by a few individualists. Saar wines can also be excellent, particularly during sunny years, for the valley is a cold one and winegrowers use oil burners to keep away untimely frosts. The best Saar wines are those of Scharshofberg.

The Ruwer (not to be confused with the Ruhr) is only a rivulet, but two of its wine districts have a deserved international reputation: the Eitelsbacher Jarthäuserhofberg, and the Maximin-Grünhause. The latter is owned by Andreas von Schubert, whose hobby is traveling and collecting antiques. It has light wines, sometimes naturally sparkling and sometimes too acid to be marketed without being sugared. But on a good, sunny year, the Ruwer wines can rival the best of the Moselles, yet remain characteristically light and subtle. Herr von Schubert still keeps his 60-odd acres of vineyard divided into four parcels that have existed ever since the Benedictines of St. Maximin made wine here. The top of the hill, with the best exposure to the sun and yielding the best wine, was reserved for the abbots, and is still known as the Abtsberg. The parcel just

URZIG, in the Moselle Valley. *(Photo Kurt Röhrig-Roebild)*

below is the Herrenberg—its wine was traditionally reserved for the monks. Still lower down lies the Bruderberg—wine for lay brothers, while the least well-exposed part of the hill, yielding noticeably inferior wine, is called Viertelberg (the "mountain of the fourth"), the fourth reserved for paying taxes.

The city of Trier is the marketing center for Mosel-Saar-Ruwer wines. It has a huge municipal cellar which can, and occasionally does, hold 30,000 *Fuder* (or over seven million gallons) of wine. There is an annual public auction of new and old wine, and frequent gatherings of professional tasters. In competitive wine tasting, preliminary to the Federal Wine-tasting Competition, wines are given a rating. The maximum score is twenty: two points for limpidity, two for color, six for bouquet, and ten for taste.

Another German vineyard is called the Nahe, after a tributary of the Rhine. It produces as much wine as the Rheingau. Many of the wines are fair, though they are never as excellent as their better-known neighbors to the west. The best come from the volcanic and slate soil of the vineyards of Schloss Böckelheim.

In Franconia, or Franken, once an independent duchy of the Holy Roman Empire, vines were planted in the 8th century by the local wine saint, St. Kilian. Today the district has a few excellent Sylvaner, Riesling, or Müller-Thurgau white wines. They are shipped in the traditional *Bocksbeutel,* a squat, flat-sided green flagon (similar ones are used for Chilean white wines).

An excellent wine comes from the slopes of Würzburg, overlooking the Main River. It is known as *Steinwein.* The name has come to be used, indiscriminately and erroneously, to designate all *Bocksbeutel*-bottled Franconian wines. The best are from the Würzburger Stein vineyards.

There is no doubt that the thorough German winegrowers have made the best of available conditions—but the pattern seems to be changing. As in so many other countries, the younger generation does not have the traditional dedication to wine making, and many youngsters are lured to the cities. Wines are increasingly expensive to make, particularly in the best regions, where the slopes are so steep that a grape-picker must almost be an alpinist, too. The result is that the great wines become scarcer, while wines of medium quality are more abundant. However, the demand for good German wines remains at a fairly high level, so that wine making can still be profitable. The growers—and the government—are well aware that quality is of paramount importance to German viticulture, and the lovers of Hocks and Moselles can rest assured that there will be no need for them to turn over in their beds every two hours.

OFF FOR THE GRAPE HARVEST. *(Photo Chapman)*

SWITZERLAND

It has been said that Switzerland has no great wines—a statement that is debatable or should, at least, be qualified. As the majority of Swiss wines are drunk within a year or two after the harvest, it could be said that none are given a chance to attain greatness. There are several reasons for the haste. The most important is that the peaceful land of Switzerland simply does not produce enough wine to quench its thirst. Another reason is that Switzerland makes chiefly white wines, which are best drunk young. Plain white table wines, bottled in the spring after the harvest, are already in possession of all the qualities they will ever have—youth, vigor and freshness. Why wait?

This general rule is not absolutely true, however; some wines do survive and age, although the visitor is rarely likely to find them. Swiss winegrowers, like so many others, keep their best wines for themselves and for their friends.

The best-known Swiss wine, if not the most distinguished, is the Fendant, which means "splitting" and describes wine from the *canton* of Valais made from the Chasselas grape. Chasselas gaily grows all over Switzerland, and the wine, going under various names, can fit many occasions—starting as an aperitif, served in glass thimbles so tiny that drinks to friends' health can be repeated many times, and ending with the fondue, assorted cheeses melted in wine and served bubbling in an earthenware pot into which guests dip chunks of bread.

AROUND LAKE GENEVA

In the vineyards of the *canton* of Vaud, Chasselas wine become Dorin—"golden." It grows along the shore of Lake Geneva, between Geneva and Lausanne, in the towns of Begnins, Tartegnin, Mont-sur-Rolle, Bougy, Féchy, and Rolle, where the largest Swiss winehouse has a flourishing export business. Chasselas also grows beyond Lausanne toward Vevey, in Lutry, Cully, Epesses, Treytorrens, Dézaley and Saint-Saphorin, commonly called Saint-Saph'. The wines made between Geneva and Lausanne are dry and light, with a full body, while those beyond

Lausanne are fruitier. Many of them leave a slight and not unpleasant aftertaste of *pierre à fusil,* or flint, which is often found in the vineyards; it is the same taste as that found in many Chablis wines. The best wines come from Dézaley, where they are smooth and rich, with a pronounced *goût du terroir,* a pleasant flavor particular to the district. Another excellent wine comes from the Tour de Marsens.

Along the snaky road above Lake Geneva, Epesses has an interesting growth, the *Trois Soleils,* combining the rays of "three suns"—the real one that shines from above, another which reflects itself in the lake, and a third reverberating from the hot dry walls that turn the slopes into a reddish and bronze tartan. Another, even better growth comes from a vale which is said to have a moving soil—diabolical business according to old-timers, who gave the wine the name of *Braise d'Enfer,* or "hellish embers."

Vevey has the tradition of a grandiose wine festival, replete with flowered chariots and pretty girls, music contributed by known composers, decorations by famed painters and celebrated stage directors, and dances by renowned choreographers. The festival, alas, takes place only four times a century—the last time in 1955. Beyond Vevey, in the district of Chablais, the Chasselas still grows; at Yvorne it gives a rich, velvety, tender wine, while that of Aigle is fuller and with more depth.

Dôle, Switzerland's best red wine, is strong in alcohol and color, and ascends rapidly to the drinker's head. The name of Dôle was given it because the plants, a mixture of Pinot Noir and Gamay, were imported during the last century from the town of Dôle across the border in France. Sometimes the wine is made only with Pinot Noir; it is then of superior quality.

Like the Alsatians, the Swiss label their wines with the varietal name, except for a few, such as those mentioned above. They also indicate the place of origin, the name of the producer, or a brand name. Full-bodied, potent, colorful, Pinot is excellent with red meats, game and cheese. The best Pinot usually

VINTNERS' PARADE IN VEVEY.
Early 19th-century engraving.
Bomsel collection. (Photo Ph. Daniel)

SWISS VINEYARDS. Grapevines defy even the Alps. *(Photo Max F. Chiffelle)*

comes from the Valais. The northern shore of Lake Geneva also has some pleasant Pinot Noir, but here it is called Salvagnin, in memory of a now nonexistent local plant. (Salvagnin is similar to Dorin, but not to be confused with the similarly spelled but otherwise quite different Savagnin of Arbois, France.)

Vinified as rosé, Pinot becomes *Oeil de Perdrix* ("eye of the partridge") and hides its strength behind an innocuous appearance. Near Neuchatel, along the Lake of Bienne at the foot of the Jura Mountains, Pinot wine is reputed to resemble nearby Burgundy, but it is scarce.

A visit to the vineyards around Geneva is well worth while. The scenery of orderly waves of vineyards and the foaming river Rhone is charming. Cooperative wineries and a few old-fashioned growers produce a sort of Fendant called *Perle du Mandement*.

In spite of the rugged landscape of Switzerland, the vine grows nearly everywhere. In the Ticino, the Italian-speaking section in southeast Switzerland, a wine called *Nostrano* ("ours") is made from the Bondola grape, but also from Merlot, Nebbiolo, and Freisa. In the German-speaking part, around the lakes of Zurich and Constance, growers produce, from tiny patches of Pinot, a *vin du pays* called Clevener which can be slightly sparkling.

GLACIER WINES

The Valais, bordering Italy and France, is the *canton* of heroic vines. Vineyards begin near Martigny at the foot of the Mont Blanc, leap over the Rhone Valley, and, clinging to rocky slopes, climb up the mountain along a huge ladder of terrace walls that can be as high as 30 foot each. Records dating to 1313 mention three sorts of plants, the Rèze, Humagne, and a *Rouge du pays,* which are still grown. And higher up, at the improbable altitude of some 3,700 feet, there grows a white wine called the *Païen* ("Pagan"), which is simply a white Savagnin, the same grape that gives, on the other side of the Jura mountains, the distinctive "straw wines" of Château Châlon.

This is the home of the "glacier wines." A visitor who has made the effort to reach Grimentz, one of the highest villages in the Valais, may, as a reward, be taken to visit the communal cellar and taste its wine, which is quite an adventure. As soon as the vat has completed fermentation, the wines (made with the Rèze, and now increasingly with the Arvine grape plants) are racked into casks of larchwood, and spend the winter in the town of Sierre. After the snow has melted, they are loaded on chariots and travel to Grimentz, some 5,000 feet up in the Anniviers Valley. Nearby, in the snow-covered slopes of the Becs de Bossons, three miles from a glacier, there are natural cellars where wines are put to age in wooden vats that are practically never emptied: the new harvest goes to fill up the vats, partly emptied after the annual sales, and becomes mixed with wine up to twenty and thirty years old (the older the wine, of course, the less of its remains in this blend) and with wines of recent years. The new wine becomes "maderized"—taking on the color and characteristics of Madeira wine as a result of oxidation; it then develops a very peculiar taste, somewhat reminiscent of turpentine, though not unpleasant, and typical of a true glacier wine.

NEAR VEVEY, vineyards are reflected in Lake Leman. *(Photo Dorozynski)*

Below, on the right bank of the Rhone, at an altitude of between 1,200 and 2,500 feet, there is a long ribbon of vineyards running for thirty miles, producing some Fendant, some Dôle, and some wines from Pinot Noir and Gamay plants. Then there is a wine called Johannisberg—no relation whatever to the famed castle on the Rhine; it is made with the Sylvaner grape, which gives so much ordinary table wine in Germany and Alsace. The Swiss Johannisberg is among the nicest wines of the Valais, richer and with more finesse than the Fendant.

Sion, the picturesque capital of Valais, has a cooperative (Provins) which has permitted growers to survive by processing and blending wines from parcels of vineyard so tiny they would otherwise be uneconomical. Two plants give a white wine of class which must be left to age for a few years: one is Amigne, which is the *Vitis aminea* of Roman origin, and the other Arvine, of unknown origin but grown here since time immemorial. A few wines of Riesling, Malvoisie, Chenin Blanc, Syrah, Pinot Chardonnay, Aligoté, and Roussette de Savoie (here known as Altesse) have been introduced by the late Dr. Wuilloud, known as the father and apostle of Valaisan wines. He gave the impetus for the creation of the *Académie Suisse du Vin,* which has assumed the task of giving Switzerland its viticultural legislation and is trying to enact a system of *appellation contrôlées*.

These white Swiss wines are eminently suited to accompany chestnuts; or smoked meat from the Grison mountains and from the Valais, which is sliced paper-thin and fluffs high on the plate; or the raclette of cheese melted in front of a fire; or the fondue.

And some Swiss wine can profitably be taken home by the wise traveler who disregards the silly story that wines of the Valais can only be drunk up there in the mountains. They are good travelers and excellent traveling companions.

STREET SIGN in pewter, 1830. Swiss. *(Photo Musée de l'Homme, Paris)*

ARRIVAL OF THE NEW WINE, "Heurige." *(Photo Chapman)*

AUSTRIA

If hardly any Austrian wines are known outside Austria, it does not mean that none are made, for Austria produces about 4 million gallons annually. Many are excellent, most of the others good to fair, nearly all white and light. Few ever cross the border, because the Austrians are great wine lovers themselves and consume an annual average of seven or eight gallons per capita.

The best and most typical Austrian wines are white, and should not be made to stand comparison with any others. Austrian wines, like Austrian people, have a humorous, hospitable, fresh disposition that is entirely their own. Many who have tasted Austrian wines prefer them as a daily fare to German growths—against which they are most frequently, and many think erroneously, compared. Austrian wines do not have, and would not care to have, the dignity, solemnity, seriousness and distinction of the great Teutonic growths. They would rather remain brash, fresh, feather-light and, now and then, irreverently sparkling.

Gay Vienna is the world's only major metropolis where consequential amounts of wine are made within the city limits. These Grinzinger, Nassberger, and Wiener wines are usually served fresh-made, under the name of *Heurige* (young wine), in countless outdoor cafes and underground cellars. The wine frequently has no label, and is served from green half-gallon bottles, which are kept cool in bucketsful of water.

The presence of *Heurige* wines in cafes and country inns is heralded by a pine branch or a straw marker hanging over the door. Although varying from vineyard to vineyard, they usually have at least a few common characteristics. They are extremely low in alcohol—around 8 percent, sometimes less; they have a light, almost white color, often milky with bubbles, and a fresh and evanescent flavor. Only a small amount is sold in bottles—standard Alsatian-type bottles, green or brown, which show the district name and sometimes the name of a vineyard and grape.

Forty miles east of the Gay City, and visible on a clear day from its famed giant ferris wheel or the revolving tower restaurant in the Prater, are the hills of the Wachau, along the Danube. Here in the picturesque village of Dürnstein, with its cobbled streets and extraordinary Baroque houses, Richard the Lion-Hearted was imprisoned on his way back from one of the Crusades. A steep climb up between the vineyards leads to the ruins of a castle, once a stronghold against the Turks, with a view over the Danube. The Wachau produces wines from Riesling, Gewürztraminer, Sylvaner, and Müller-Thurgau grapes, and from a few local varieties, such as Veltliner (an excellent white grape, whose wine is not unlike a Traminer) or Rotgipfler.

Austrian vineyards are considerably more to the south than the great German districts, and the wines are less acid. *Spätlese* are nevertheless produced, to give full-bodied, fragrant wines which are among the few in Austria that gain from aging. Kremser, Leibner, and Dürnsteiner *Spätlese* are elegant and cosmopolitan, and the *Auslese* have a sweetness and a wealth that is not cloying.

Some thirty miles south of Vienna, around the pretty village of Gumpoldskirchen, lie the best-known vineyards in Austria, rising steeply to wood-topped hills or stretched out on the plain below. Gumpoldskirchener wines are fresh and some have an extraordinary bouquet. The name of the grape is usually given, though sometimes the wines are blends. When the wines come from the best known vineyards—such as the Marienthaler—this is also usually indicated. Of course, the beloved *Heurige* is made here too, and served in "Heurige cafés."

White wines of a similar nature, though never as excellent, are made near Baden and Vöslau. Sweeter wines made near Rust, in the Burgenland district, are highly praised.

TYROLESE VINEYARD. *(Photo Kurt Röhrig-Roebild)*

SPAIN

THE SHERRIES

"A good sherris-sack hath a two-fold operation in it. It ascends me into the brain; dries me there all the foolish and dull and crudy vapours which environ it; make it apprehensive, quick, forgetive, full of nimble fiery and delectable shapes; which, delivered o'er to the voice, the tongue, which is the birth, becomes excellent wit. The second property of your excellent sherris is, the warming of the blood; which, before cold and settled, left the liver white and pale, which is the badge of pusillanimity and cowardice; but the sherris warms it and makes it course from the inwards to the parts extreme . . ."

If any wine can claim to be the world's best, it must be this Sherry described by Falstaff, who added, "If I had a thousand sons, the first humane principle I would teach them should be, to forswear thin potations and to addict themselves to sack." Generations of Anglo-Saxons have, indeed, addicted themselves to Sack, which is produced near the small Spanish city of Jerez (pronounced Haireth) de la Frontera, some 80 miles northwest of Gibraltar, near Cádiz in the province of Andalusia. "Sack" as a name comes from the Spanish *saccar*, meaning "to transport" or "to carry," and indicates that the wine is made almost exclusively for export.

Produced from the Listan (or white Palomino) and half a dozen other grape varieties, Sherry comes from a soil called *albera*, a harsh mixture of clay and chalk. To cultivate this soil, a Spaniard once wrote, is like "working in powdered marble or in petrified snow." Under a heat of 130 degrees, the sun hardens the earth and reflects the sunshine like a giant mirror, and the vine's roots must dig deep in search of humidity. It is this soil which gives the best Finos, the lightest and generally driest Sherries. The nearby *barros*, or clay soil, is more fertile but its wine is less distinguished, while *arenas*, or sand, gives a high yield but with far less distinction.

The care bestowed upon the vineyards is almost incredible. A Spanish viticulturist once advised that the vine's thirst should be quenched with rose water, to develop aroma; another recommended that the berries should be protected against bees and wasps by being painted with a thin dewlike coat of virgin olive oil. Without going to extremes, each plant nevertheless requires 24 visits every year. Today, the tractor can sometimes be seen, but most of the cultivation is still done with a plow pulled by half a dozen mules, or by hand with the spade, the pickax, the sickle, or the *navaja*—a huge, hatchetlike knife.

At the beginning of September there is a first harvest of early-ripening grapes, to be followed later by a second when the remaining grapes are ready. The harvest is collected in boxes or baskets woven of olive boughs, and brought on donkey-back or in oxcarts, to be left outdoors to take the sun for a few hours or a day or two, in the ancient Greek fashion. Then the grapes are thrown into *lagares* (the *lagar* is a sort of wooden press trough) to be prepared for the *pisa*—treading by foot. Young girls wearing dresses of ancient and traditional design spread the grapes evenly, and four young boys, the *pisadores*, wearing short pants and cleated shoes, start a sort of traditional, Dionysian dance, while the village bells ring and white doves are released, each carrying a little bacchic poem tied to its foot, to announce the beginning of the harvest festival.

Andalusian winehouses—called *bodegas*—are not underground cellars, but buildings leaning on columns, white with chalk, covered with a Flemish-style frame and Roman tiles, and mostly owned by English firms. There are some 700 in Jerez, 200 in Puerto de Santa Maria and 300 in San Lúcar. They are so huge that Théophile Gautier compared them to cathedrals, and Henry Ford to navy yards. Light is admitted through small round windows screened with dried grass, and there are barrels everywhere—inside and outside. They are all 150-gallon casks made of oakwood, in which Sherry wines will be blended and aged.

By December, the new wine has fermented and is clear enough for tasting. It is dry and of a pale color, but an experienced taster will immediately recognize it as belonging to one of two principal types: either a dry and light wine, or a heavier and sweeter one, with fuller body. If light and very dry, it will have on its surface a thin layer of a yeast, called *flor*, a native of Jerez. This cask is destined to become a Fino, and is chalked with a sign, the *palma* (or palm leaf), which looks like a Y. The sign may be repeated two or three times, to put emphasis on the excellence of the wine. A wine somewhat less dry will be Amontillado (named for the village of Montilla, south of Córdoba, whose wines used to be legally sold as Sherries). The wine is drawn off into a new barrel, and fortified with grape brandy to reach an alcoholic content of 15½ percent.

If there is no *flor* or too little of it, the wine is heavier, with less bouquet, and is chalked with a diagonal mark, the *raya*, to be blended into another type of sherry, such as the Gordo. Those wines found to combine the qualities of both types are marked with a *raya y punto* (a slash and a dot) and, in between, there are varied combinations of dashes and dots. If wine is not found to be up to standard, it is scratched off (three dashes) and goes to make wine brandy—either to be used to fortify Sherry or to be sold as is.

Thus having its future outlined in chalk, the wine goes into a part of the *bodega* called the *criadera* (the cradle), which is a nursery where it will remain for one or two years. The Sherry wine itself will be born in the *solera*, an assemblage of butts in which the wines are mixed progressively as they mature.

Solera

The *solera*, from the Spanish word *suelo*, meaning the soil or foundation, consists of a set of barrels superimposed usually in three or four tiers. A large *bodega* in Jerez will have at least a dozen *soleras*, each one made up of as many as 100 barrels, lined up in rows under the roof.

THE OLDEST CELLARS in Jerez de la Frontera
belong to Gonzalès and Byass. *(Photo Brassaï)*

IN THE PEDRO DOMECQ CELLAR, old kegs still bear the names of Fox, Nelson and Napoleon. *(Photo Brassaï)*

KEG bearing the signature of bull-fighter Juan Belmonte, in the Gonzalès and Byass cellar in Jerez de la Frontera. *(Photo Brassaï)*

The *solera* process is continuous. New wine arriving into the top tier of barrels is progressively blended with wines in the barrels below, finally emerging at the bottom to go into the bottle. It is started with a particularly good wine, corresponding to a type that the *bodega* seeks to reproduce from year to year. Most of the going *soleras* were started scores of years ago, so that the usual process consists of drawing wine from the bottom tier for bottling, refilling those barrels from the ones just above, and so on. The new wine, from the *criadera*, is then poured to fill in the empty space on top, and the process continues. Thus the wine that is bottled can be a mixture of as many as 100 different vintages.

Although there are variants in the number of tiers and their disposition, the *flor* yeast *(mycoderma vini)* is always the key factor in giving the wine its particular flavor. The cellar master, of course, is responsible for the final result, since he selects the wines to be mixed in order to achieve the desired standard.

Clearly there can be no vintage Sherry (though there are ways of misleading the customers into believing there is). The only valid indication having to do with time is usually found on the cork, which can be marked with the year of bottling. At any rate, there is no good reason for Sherry "vintages" to be indicated, for ideally the *solera* blend should be unchanged.

The color is usually the best indication of a Sherry. The *muy palido,* very pale, is dry, such as, for instance, the Fino and the Palmas, with 15 to 17 percent alcohol, and an aroma of fresh flowers and of almond. *Palido* is the slightly darker shade of the Amontillado, more concentrated than the Fino. The more

VINEYARDS NEAR TARRAGONA. *(Photo Bildarchiv Irmer)*

full-bodied and generous, though less distinguished, wines have had, and still have, their partisans, who have ranged from Henry VIII of England and Francis Drake to Nicholas II, the last tsar of Russia. The color is *oro,* or gold, for Oloroso, a wine of medium sweetness; *oscuro* or dark, for the sweeter, so-called Cream Sherries and for some Olorosos on the sweet, more full-bodied, side. And it is *muy oscuro,* or very dark, for the so-called Brown or East Indian sherries that used to travel to India and back to help them age. In between these types there are variations—notably, the luscious Amorosos ("in love"), sweet and rare essences, the best of which are made of the Pedro Ximines (PX) grape plant.

Also made from the PX, picked very ripe and left two to three weeks under the sun, are the Dulce sherries: the wine is almost a liqueur, often used to sweeten other Sherries.

Then there is Manzanilla, which is legally called a Sherry, but is, in fact, somewhat different. Light, pale, and extremely dry, Manzanilla is made just outside the Sherry district, near the small town of Sanlúcar de Barrameda, on the ocean, at the mouth of the Guadalquivir River, where Columbus set sail on his third voyage to the New World. In addition, there are many subtle varieties between Sherries produced by one *bodega* and the other. True Sherry is still primarily an export wine (often shipped in oak casks) but the amount of Sherry made in Spain is engulfed in a tide of imitations made in South Africa, Australia, and the United States (which alone produces four times as much "Sherry" as Spain). The method by which sherry is made in Jerez can, however, give good results elsewhere, although it is not always used for Sherry imitations.

In Spain itself, the production of Sherry (with which many erroneously identify all Spanish wines) amounts to about 10 million gallons a year—only a drop in the ocean of 500 million or so gallons of other wines produced in Spain.

SOME SPANISH VINEYARDS

The yield of Spanish vineyards is relatively low, and the wine-making methods are often archaic. Nevertheless, many of the wines are likely to pleasantly surprise the visitor.

The king of Western Andalusian wines is Montilla, produced near the town of Córdoba; it has a pale greenish transparency, a gentle bitterness and, Andalusians claim, a penetrating flavor of thyme, dianthus, almond, and the air of the Sierra del Macho. Along the coast, east of Gibraltar, in the province of Málaga, is made the sweet, brown fortified *solera* wine Málaga; it can easily be distinguished from its many imitators because it never cloys. Some Málagas are sweetened by *arrope,* a grape concentrate, but the best are made with the sweet juice of Pedro Ximenes grapes dried in the sun. There are a dozen different types of Málaga, with color ranging from dark brown to gold-red and pale yellow.

North of Andalusia, toward Madrid, there stretches a wide, bare and rolling country, with hardly any trees or houses. It is *la Mancha,* the scene of Don Quixote's tribulations, the desert and the heart of Spain, with scorching days and chilly nights, and a dark soil of stone or clay. It forms a part of the Valdepeñas ("valley of stones"), an important wine-producing district. The huge area produces yearly some 125 million gallons of wine, some of which is good light and light-colored table wine served in Madrid; another type is full-bodied, amber-colored white wine, much of which is exported to Switzerland and Germany, often to be blended.

Due east is another important wine-growing region—the second in Spain in size, and third by volume of production: the Levant. Along the coast are orange groves, rice fields, palm trees, gardens and vegetable patches, but inland, from the Segura River to the Ebro, more than 200 miles as the crow flies, three provinces make and export red, white, and rosé wines: Murcia, with its towers and palm trees; Alicante, gentle and luminescent; and Valencia, an active modern city behind fortified gates.

The Swiss buy much of the region's rosé wine, made with the Monastrelli grape. Near Alicante, the dry weather turns this wine cherry color; other red wines are chiefly reserved for blending. The best-known wine of the region is produced from a "Roman" Muscat, a grape which can be exported as a table grape or made into wine with the characteristic muscat aroma.

Northward beyond the Ebro is the heart of the Catalonian provinces, with rocky slopes scorched by the sun, and with vineyards rolling down to the seashore. Here the walls of the ancient Roman city of Tarragona still seem to stand guard over the land which attracted the monks of St. Bruno, who came from France in 1163 and built the first Spanish charterhouse, Scala Dei. Some of the wines made with the Carignan and Grenache grapes are called the wines of

Priorato. Strong in alcohol, the Priorato red wines are particularly remarkable when they became *rancio;* in this process, after being aged in casks, they are kept for two or three years in demijohns buried in the ground, during which they undergo maderization, which gives them a beautiful fawn color. The winter cold meanwhile favors the precipitation of sediment. The wines are then stored in barrels, mixed equally with wines of the preceding year.

The young Catalonian red wine has a deep dark color and is often used in making *Sangria,* a refreshing brew. To make *Sangria,* add two to three lumps of sugar to a jugful of the wine, plus a few slices of lemon and orange, sliced peaches, and a pinch of nutmeg if you wish; add also ice cubes if they are handy, and plain or soda water to your taste.

Between Tarragona and Barcelona lies the region of Panadés, birthplace of the best Spanish sparkling wines, made in the traditional Champagne fashion, and called Xampán (pronounced like the French *Champagne*). Villafranca del Panadés is the site of a wine museum, with a collection of ancient wine vessels.

RIOJA WINES

The best Spanish table wines are produced south of the Basque country, in the Rioja region, with its four provinces of Logroño, Alava, Navarra and Burgos. This is a mountainous region, with long winters that leave the snow along the valleys as late as April, with temperate springs, and a month of May not unlikely to be stormy, and with cool, sometimes humid summers. Vineyards grow fairly high along the mountain slopes, and wines are fresh, well balanced, usually lighter and drier than Bordeaux wines, which they somewhat resemble.

Wine has been known here at least since Roman times, but the vineyards really advanced after the devastating vine pest, the phylloxera, struck the vineyards of the Bordeaux region toward the end of last century. Many French winegrowers and their families abandoned their ravaged vineyards and moved to Haro and

WINE CELLAR in a Madrid suburb. Engraving from *"L'Univers." Mella collection.*

Logroño, the center of the Rioja district, bringing with them their ancestral know-how. The phylloxera struck Spain a few years later, but the vineyards were quickly replanted. The Rioja was the first Spanish wine-growing area to form a protective "regulating district"—of which there are today seventeen, representing Spanish equivalents of French *appellations contrôlées*.

Most of the plants used are native: Garnacha (the Grenache so widely used in other countries), Tempranillo, Mazuela, Graciano, Miguel del Arco and Monastrelli. They yield a wine with usually 11 to 14 percent alcohol, light, and often distinguished, with a faint and pleasant earthy flavor. Generally there are no vineyard names on labels, and the best way to find good wines is to rely on the name of a reputable producer or merchant, such as the *Compañia Vinicola del Norte de España* (C.V.N.E.), the *Marqués de Riscal, Federico Paternina, Bodegas Bilbainas*, and *Bodegas Franco-Españolas*. The best growths are often kept as a "reserve." Some age for many years, acquiring distinction and smoothness.

White wines are made with the Malvasia, Viura, Calgarano, Moscatel, and Torruntes grapes, but they are much less interesting; either sweet or dry, they usually contain between 11 and 12½ percent alcohol.

THE CANARY ISLANDS

There is another beverage under the Spanish flag, a jolly and volcanic little rosé, whose label is never seen at any wine merchant's. The only way to taste it is to stop off at the Canary Islands, once known as the Fortunate Isles—a paradise where spring is twelve months long, and where fruit, flowers, and colors are countless and unforgettable.

The most beguiling wine comes from the village of Taganana on Tenerife Island. It is a dry rosé, with a strong whiff of earth designed to bring a body back to life. It is not made to age; significantly enough, when it is bottled, the cork is often left protruding, so that no time needs to be wasted looking for a corkscrew. Many of the vines from which this rosé is made are allowed to crawl along the soil under the tropical sun most of the year; in June the farmer drives stakes into the ground to prop the plant up to receive more sun.

Canary Sack, once known the world around, is now extinct. But the best wine of the archipelago today was also present at the European courts in the 17th and 18th centuries. It is the famed Malvasia (or Malmsey), a white wine with the flavor of Turkish tobacco, burning grass, and raisins. The first plants were imported from Crete by Prince Henry the Navigator. The wine is dry and smooth but full and strong in alcohol—16 to 18 percent—and of a beautiful topaz color.

Malvasia can age well—but then, few wines get a chance to age in the Canaries because they are so promptly consumed. There are no cellars; the wines are kept in a *salon*, made of stone or cement, of a size in proportion to the wealth of the owner. And the corks are not driven deeply into the bottle necks, so that that they can be easily and rapidly removed.

Canary wines have disappeared from the world market—but have left descendants behind. Their offspring have settled in Chile, Peru, and Argentina.

PORT LABELS of the last century. *(Photos Bernard Jourdes)*

PORTUGAL

Portugal produces dozens of excellent table wines, of which only a few are internationally known. And even these are completely overshadowed by a giant, the world's most famous dessert wine—Port.

Port and similar fortified wines have been made in Portugal for more than 500 years. It is believed that Port first reached England in 1678, when the two sons of a wine merchant in London visited a monastery in the Douro River valley and were served some very sweet wine to which grape brandy had been added. The visitors took a few bottles back home and these met with such success that the merchant sent his sons back to Portugal for more.

Twenty-five years later the English signed the Treaty of Methuen with Portugal, reducing the import duty on Portuguese wines to seven pounds a tun (a large cask), while French and German competitors were taxed at the rate of 55 pounds. Port wine was then not only fashionable in England but also inexpensive. And as unfriendly relations prevailed between France and Great Britain, Port also became the patriotic drink.

> *Be sometimes to your country true,*
> *Have once the public good in view,*
> *Bravely despise Champagne at Court*
> *And chose to dine at home with Port—*

wrote Jonathan Swift. By 1790, England was importing over 100 million gallons of Port a year. Ever since, Port has remained a part of the British way of life.

Made in a strictly controlled district, Port represents only between 1 and 3 percent of Portuguese wine production. But it is one of the world's most imitated wines. The Soviet Union makes hundreds of thousands of gallons of "Portvein" a year, and California produces five to ten times more "Port" than Portugal itself—though Portugal produces nearly twice as much wine as California.

True Port is born in a deep valley which cuts across chalky and stony mountains covered with pine and chestnut trees in the northern part of Portugal. The

TRANSPORTING PORT TO THE Villa Nova De Gaya warehouses, where it will age. *(Photo Seto)*

Douro Valley runs from the Spanish border to the Atlantic coast, and its sides are scarred with scores of smaller valleys and craggy ruts. One of Western Europe's most rugged regions, the valley is bitterly cold and foggy in winter, and racked by violent storms in spring. Summers are hot and dry.

Planting and upkeep of the vineyards represents the patient, relentless work of men who have transformed more than 100 miles of the valley's walls into stone terraces lined with traditional vine plants—Tinta Francisca, Mourisco de Semente, Bastardo, Malvasia Tinta, Rufeta and a few others.

Throughout most of the year life in the Douro Valley proceeds at a slow pace. But in the last days of September, there comes an awakening. Recruiters *(radagores)* trek through the villages, assembling teams of grape-pickers who march along the dusty roads to the sound of a *fado*—a type of plaintive, Portuguese folk music—rendered on the accordion or the mouth organ. Landowners make ready their cellars, clean up their barrels, load casks of brandy onto oxcarts, and repair to their *quintas*—their vineyards along the Douro.

During the harvest men wear their oldest patched-up clothes and cover themselves with sacks to carry the tall brimming baskets from which the sweet juice runs down their backs. The baskets, weighing up to 150 pounds, are held by shoulder straps and a leather strap braced around the forehead.

The grapes are thrown into *lagares,* stone or cement tanks of varying surface but all about two feet in depth, each holding 20 to 30 "pipes" of grapes (a pipe, the traditional Douro barrel, contains about 140 gallons). When the *lagares* are full, the men gaily roll their trousers up above their knees, wash feet and legs, and line up in the tanks. Arm in arm, the men march in step, back and forth over the pulpy mass of grapes. In the dim light that filters from outside, or under dull yellowish light bulbs lazily swinging from the ceiling, they go on, singing a thin, monotonous, persistent tune, until the grapes are crushed.

The fermentation is never allowed to be completed. When it has reached a point where the desired amount of sugar still remains, the fermenting juice is drawn off, leaving its sediment behind, and is introduced into a pipe which contains grape brandy, the proportion of wine to brandy being about five to one. The earlier the fermentation is interrupted with brandy, the sweeter the Port will be; and the later, the drier.

Meanwhile, around the wineries, villagers gather in a festive mood and colorful garb: girls with wide woolen skirts, white blouses, and gay scarves; men in green, brown, black, or blue pants, a wide yellow or scarlet belt, and a vest floating over an immaculate white skirt. A girl is chosen among the grape-pickers to present the owner of the vineyard with a *rama*, a gaudily decorated stick with grapes and colored strips of papers dangling from it. Fireworks break the dusk. The harvest is in—it is time now to drink, dance, and make merry.

But the wine is still far from being Port. Few growers have ageing cellars, and the pipes of new wine are loaded aboard oxcarts and transported to the Douro. There they are put aboard *rabelos*, the ancient square-sailed riverboats that will take them down to the calm, wide estuary and finally to Vila Nova de Gaia, on the river's left bank. Here the wine will rest and age under the watchful eye of the National Institute of Port Wines, the organization which controls the quality of genuine Port.

There are several types of Port. The best, labeled with its vintage year, comes from pipes selected in exceptionally good years, and never blended. This wine is bottled in its youth and ages twenty, thirty, fifty years and more. Much sediment accumulates, frequently forming a crust on the sides of the bottles, hence the wine needs to be carefully decanted. A step below these Vintage Ports are Crusted Ports, blends said to be of "vintage character."

Most Port, however, is aged in wood and bottled only when it is considered to be ready to drink. Ruby Port is a few years old, fruity and fresh. If the same wine is left to spend more time in the pipe, its color becomes lighter, acquiring a brownish tinge. It has become a Tawny, softer and rounder, more delicate than Ruby—and more expensive, too.

White Port (from Rabigato, Malvasia, Gouveio grapes) is pale yellow but becomes darker. With age, both red and white tend toward the same color. White Ports, processed exactly the same way as reds, can be excellent, but never as superb as the best from black grapes.

One of the sixteen grape varieties used to make Port can also yield a distinctive, dry table wine with an unmistakeable flowery scent. The grape is called

THE DOURO VALLEY.
(Photo Bernard Jourdes)

BARRELS OF PORT afloat on the Douro, near the city of Oporto *(Photo Bernard Jourdes)*

Bastardo, and the wine, White or Brown Bastard, enjoys some popularity in Britain, where it has been known for six centuries.

MINHO AND OTHER WINES

In contrast with the severe Douro Valley, the province of Minho, a few miles north, is a fertile garden of thousands of colorful patches, criss-crossed by springs and dotted with humble farmhouses and ancient manors, with tiny fields and green pines. On its north the Minho River forms the boundary with Spain.

Grapes here grow everywhere except in vineyards—they grow along roadsides, up the walls of farmhouses, in trellises overhead, leaping from tree to tree, sometimes as high as twenty feet overhead.

Minho's unusual wines must rightfully be listed among the world's more pleasant and lightest table wines; they are *vinhos verdes*—literally the "green

wines." Young and fresh, they are light in alcohol (8 to 10 percent), sometimes greenish in color, or pale and evanescent gold. The best are white, but some red ones are also excellent.

These wines were well known before Port became famous. They issue, it is proudly claimed, from grapes that came from ancient Greece. The principal white grapes are Azel Branco, Dourado, and Alvarino, though each one of the smaller districts in the region has its own favorite. White *vinhos verdes* are pale and lightly fruity; they are "summer wines," leaving a fresh taste in the mouth and a cool head. And without doubt they are eminently suited to bring out the qualities of Portuguese sea and river food. These have hundreds of variations made with oysters or scallops, lobster or shrimp, crab, octopus, or squid, as well as the inevitable *bacalhau* (cod prepared in various ways) or *caldeirada a franateira*, the local version of *bouillabaisse*. Red *vinhos verdes,* with their intense, lively color,

MINIATURE from a Portuguese manuscript, enlarged.

sometimes have a slight and irreverent foam, and are fresh and thirst-quenching. The basic plants are the Vinhao, Barracal, Espadeiro, and Azel Tinto.

To the east, the province of Traz os Montes, the land "beyond the mountains," produces wines that are light and inconsequential but quite pleasant. Here, once a year, there is a wine pilgrimage which requires everyone to stop and have a drink at every tavern on the way. The winner is the man who stands up the longest. On the following year, he will be umpire of the contest.

South of the Douro, the Dao River valley stretches down toward Lisbon, but never reaches it. This small, mountainous wine district produces rich, strong wines, tasting of grape and sunshine, that some say have a resemblance to French Burgundies. The vineyards are centered around the ancient ecclesiastical, walled city of Viseu, where, probably for the first time in history, an American came to figure in a biblical scene. Scenes of the life and passion of Christ in the cathedral, painted by Jorge Alfonso shortly after the discovery of Brazil, show a feathered Indian instead of an African king in the scene of the Adoration of the Magi.

The Dao Valley wines are made chiefly from the same grapes as those used for Port. The white and black grapes are frequently mixed and the wine rapidly racked after pressing, giving it a brilliant, transparent, ruby color. Dao wines often have a smooth, velvety quality, but they are nevertheless quite heady, having a rather high alcoholic content—upward of 12 percent.

Further south along the coast lies Estramadura, with its bat-winged windmills, and famed resort cities. Here, between fashionable Estoril and historic Sintra, the Colares River region produces wine from vines growing on sand dunes. The vines must frequently be protected from the wind with walls of reed or with branches of pine trees held together with wicker. The most widespread plant in the district is the black Ramisco, and its wine, even when blended, has a ruby color that acquires a brown tinge with age, and a smooth and heady aroma with a whiff of almond and of violets, sand, and the ocean.

Almost directly south of Lisbon, across the estuary of the Tagus (now spanned by the huge Salazar bridge) and the Arrábida peninsula, lies the town of Setubal, a dense city squeezed among beautiful orange groves, vineyards, and magnificent beaches. The town, said to have been founded by the Patriarch Tubal, son of Japheth and grandson of Noah, is the center of an important wine district known for its excellent golden Muscat wine, the Moscatel de Setubal.

The Setubal region also produces a Moscatel Roxo wine—strong and opaque, without its white brother's distinction. Nearby, a recent addition to Portuguese wine production has been a lightly sparkling and prettily pink wine, which has reached the United States in jugs under the name of Lancer's.

MADEIRA

Far out in the Atlantic, west from Morocco, lies Madeira. Falstaff sold his soul to the devil "on Good Friday last for a cup of Madeira and a cold capon's leg," and many Englishmen today might sell their soul likewise, if their Madeira wine was not so readily available.

Grown in a subtropical climate, Madeira wines are sweet and fortified. They are aged, like Sherry, in *soleras,* after having been kept for several months in *estufas,* rooms especially designed to keep them at a constantly high temperature. There is a great variety of Madeiras, ranging in strength from 17 to 21 percent alcohol, usually named after the grape from which they have been made. The driest and best is the Sercial, with an unmistakeable and inimitable bouquet. Other Madeiras range from pale straw color (known as "rainwater," after an American vintner, Rainwater Habisham, who developed a special method of processing this wine) to deep gold Malmsey. The latter is made from the Malvasia grape, presumed to have originated in Greece. These wines make excellent aperitif and dessert wines.

Solera wines, of course, have no vintage, and hints of great age in Madeira wines should be treated with caution. For example, mention on a label that a Madeira was "laid out two years after the battle of Waterloo" may be misleading, since a *solera* started in that year (1817) would contain, a dozen years later, only infinitesimal amounts of the original.

A MADEIRA LABEL, 19th century. *(Photo Chapman)*

THE TASTE. Colored engraving by Abraham Bosse.
Musée des Beaux Arts, Tours. (Photo Arsicaud)

BELGIUM AND LUXEMBOURG

Thin people are rare in Belgium and Luxembourg. Tables here are usually loaded with a variety of foods, often evoking the luxurious feasts of the Flemish masters. The beer is excellent too, but it cannot compete with good wine in generating harmony and good humor at a reunion. This explains why the Belgians rank as the Number One drinkers of Burgundy outside of France.

Nothing is left of the rich vineyards that grew in medieval times around the abbeys, convents, and castles surrounding Liège, Namur, Louvain, Bruges, Ghent, and Brussels. But Belgians are tenacious, even stubborn: today their vineyards grow in hothouses, with coal replacing sunshine. The wine cooperative of Isca manages to produce about 100,000 gallons á year. More recently, an even more interesting experiment consisted of keeping the hothouse temperature at the proper level with the help of nuclear energy. Such prowess and persistence should suggest that one look upon Belgian wines with some respect, and grant them at least an indulgent tasting, keeping in mind they do not claim to be great.

The people of the Grand Duchy of Luxembourg also like a high-calory diet, which includes suckling pig, haunch of hare, and backbone of pork. Luxembourgers grow their own wine, ·from some 2,000 acres of vineyards along the northern bank of the Moselle. Most are made with Riesling; they are fresh, light, sometimes slightly sparkling, fairly acid and tart, and hence particularly pleasant in the summer. The town of origin and the grape are indicated on the label. The best-known districts are Grechen, Ehnen, Ahn, Maehtum, Remich, Wintringen, Wormeldange, Grevenmacher and Wasserbillig.

Tel. 23698

14, Rue El Shohadaa
ALEXANDRIE

★

الـنبيـذ العربي
THE ARAB WINE

تليفون ٢٣٦٩٨

١٤ شارع الشهداء
الاسكندرية

★

VINS GIANACLIS — أنبــذة جاناكليس

	PRIX COURANTS		P.T.		الأسعار الجارية
rosé	Rubis d'Egypte	la Dz.	438	٤٣٨	روبي مصري
	Rubis d'Egypte	½	264	٢٦٤	روبي مصري ½
red	Omar Khayam		456	٤٥٦	عمر الخيام
	Omar Khayam	½	264	٢٦٤	عمر الخيام ½
	Castel Nestor		384	٣٨٤	كاستل نستور
	Castel Nestor	½	222	٢٢٢	كاستل نستور ½
	Reine Cléopatre		384	٣٨٤	الملكة كليوباترة
	Reine Cléopatre	½	222	٢٢٢	الملكة كليوباترة ½
	Village Gianaclis × white sec		342	٣٤٢	قرية جاناكليس
	Village Gianaclis	½	192	١٩٢	قرية جاناكليس ½
	Cru des Ptolémées		348	٣٤٨	بطالمة
	Cru des Ptolémées	½	168	١٦٨	بطالمة ½
red	Château Gianaclis		288	٢٨٨	قصر جاناكليس
	Château Gianaclis	½	168	١٦٨	قصر جاناكليس ½
	Clos Mariout white sec		252	٢٥٢	كلو مريوط
red	Clos Matamir		252	٢٥٢	كلو مطامير
	Gianaclis Abyad		216	٢١٦	جاناكليس ابيض
	Gianaclis Abyad	½	120	١٢٠	جاناكليس ابيض ½
red	Gianaclis Ahmar		216	٢١٦	جاناكليس احمر
	Gianaclis Ahmar	½	120	١٢٠	جاناكليس احمر ½
	Retsina Rosée		258	٢٥٨	رتسينا حمرة
	Retsina		240	٢٤٠	رتسينا
	Vin de Messe Rouge Doux (Commandaria)		336	٣٣٦	نبيذ قداس أحمر حلو (كومنداريا)
	Vin de Messe Blanc Sec		264	٢٦٤	نبيذ قداس
✗	Kokkinelli		252	٢٥٢	كوكنلي
	Kokkinelli	½	150	١٥٠	كوكنلي ½
	Brandy		1122	١١٢٢	براندي
	Brandy	½	600	٦٠٠	براندي ½
	Ouzo		1056	١٠٥٦	اوزو
	Zibib		780	٧٨٠	زبيب
—	Porto Alex		594	٥٩٤	بورتو الكس
—	Vermouth		486	٤٨٦	فرموت
—	Muscat d'Egypte		486	٤٨٦	موسكات مصري
	Muscat		366	٣٦٦	موسكات

VIN EN VRAC — نبيذ سايب

Blanc	le Litre	21,5	٢١,٥		ابيض
Rouge		21,5	٢١,٥		احمر
Kokkinelli		26,5	٢٦,٥		كوكنلي

Couronne d'or (mousseux)

North Africa, Greece and the Near East

ALGERIA
TUNISIA
MOROCCO
GREECE
ISRAEL
LEBANON, SYRIA, JORDAN
EGYPT
TURKEY
IRAN

YOUTH holding a bunch of grapes.
Egyptian bas-relief. *Berlin Museum.* (*Photo Boudot-Lamotte*)

Preceding page: THE WINE LIST of an Egyptian merchant in Alexandria.

NORTH AFRICA

Throughout history wine has been made by those who used it—either as a sacrament or as a beverage, or both. The long and exacting work of selecting just the right vines for any given soil and weather conditions, and the years of trial and error necessary before the ideal methods of vinification are established, could be tolerated, one would think, only by those who regularly sample their product with love and dispassion (strange bedfellows, perhaps, but essential to the development of fine wine). All the more remarkable, then, are the vast amounts of wines that emanate from North Africa, where they are made chiefly by Moslems, to whom the fermented juice of the grape is forbidden.

It is true that wine grapes—as opposed to the many excellent table and raisin varieties cultivated by Moslems—were introduced to Africa by Europeans. Nonetheless, Moslems almost exclusively tend the vineyards today, make the wines, and ship them mostly to their thirsty European neighbors. And although a vast percentage of the wine is bulk-produced, of high alcohol content and mediocre quality, there are some growths that are extraordinarily good. That wines of good quality can be made by those who do not partake of them is another paradox in the amazing kingdom of wine.

ALGERIA

Of the 750 or so million gallons of wine that traveled through the world in 1962, nearly half went from Algeria to France. This vast amount of wine—in fleets of tankers and tank trucks and miles-long strings of railroad cars—represented nine tenths of the produce of the youthful state of Algeria, which that year regained its independence from France.

The first French settlers, who came in the 1830's, grew only tobacco and corn, and one of the early governors, Marshall Thomas Bugeaud de la Piconnerie, opposed wine growing in the colony. By 1850, only about 20,000 acres of vineyards existed in Algeria. The impetus that established the wine grape in Algeria came toward the end of the century, brought about by two developments. The first was the invasion of southern France by the devastating vine louse, the phylloxera, which forced the French to look elsewhere for a source of wine. The second was the influx of French refugees from Alsace-Lorraine, then part of Germany. The French government encouraged the refugees to take land concessions in Algeria either free or at an advantageous price.

The prosperity of the settlers grew continuously from 1880 on. Vines were planted chiefly near Algiers and Oran; the *bled*—properly the vast, landlocked interior of the country, but a name soon synonymous with any isolated, deserted area—was given life. An 800-mile strip of uncultivated seashore, started to look more European. So did arid mountain slopes, fields of esparto grass, and the wildest corners of the Tell, the rugged hinterland flanking the Mediterranean.

Today the Algerian vineyards cover an area of some 850,000 acres divided into 32,000 properties. Most of the growers own less than 25 acres, and the entire family works on the vines. The bulk of the product is either ordinary table wine or wine for blending. But each of the three principal wine-growing districts in Algeria can boast of a few high quality wines.

One of the districts, that of Oran, produces at an altitude of 2,000 feet two of the great Algerian growths. Mascara, best known of all, has a full body and a strong bouquet which helps it age extremely well. It is a strong wine, with up to 15½ percent of alcohol. Tlemcen is Mascara's younger brother, less full-bodied and less strong. Both the Tlemcen and the Mascara districts, as well as those of Sidi Bel Abbès, Aïn Témouchent, Oran and Mostaganem, produce mostly wines sold for blending.

In the region of Algiers, the districts of Miliana, Médéa and Bouïra produce fine mountain wines, with a strong and pleasant perfume, and brilliant color. The slopes of the Sahel, bordering the Mitidja plain, and the Dahra, as well as the coastal area at Ténès, give fruity wine, both for blending and bottling.

Between Algiers and Tunisia, around Bougie, Djidjelli, Phillippeville and Bône, the third wine-growing region of Algeria produces chiefly light wine, with a few fine table wines on the slopes.

Algerian red wines have an alcoholic content ranging from 11 to 15 percent. Those growing on the plains are made chiefly with Aramon, a high-yield plant, Cinsault and Alicante Bouschet—these three plants make the strongest wines. Finer red wines are made from Cabernet and Pinot grapes, rosés with Cinsault, Grenache and Clairette. A few white wines are made with the Clairette, Ugni Blanc and Aligoté grapes.

Wines are not sold under the varietal names, but usually under fancy labels, such as Hoggar, Kebir, Valpierre—or under the name of the producing community, such as Mazagran, Médéa, Targui, Mascara, Mansoura, Hammam-bou-Hadjar, or Aïn-Kial.

Since independence, most of the French growers have left the country, and many large estates have been taken over by the new government and subdivided. The future of the Algerian vineyards depends on the willingness and ability of nondrinkers to make wine. And it is certain that the Algerian government will do everything it can to preserve one of its principal sources of income.

ALGERIAN WINE LABEL, 19th century. *(Photo Federico Arborio Mella)*

VINEYARDS on the Mitidja plain, Algeria. *(Afrique photo)*

TUNISIA

Tunisian wines are being introduced to the world. In recent years, "Tunisia Houses" have been established in New York, Hamburg, Hanover, Amsterdam, London, and Brussels, to let people know that Tunisia welcomes tourists, and to acquaint potential customers with Tunisian products.

Already the wines have found some takers, and their renown is likely to increase, for many growths are excellent, coming from fine plants grown on suitable soil, receiving abundant sunshine and sufficient water. Modern wine-making methods have been developed by French and Italian settlers, and prices are more than reasonable. But the Tunisian wine list is limited, with labels usually bearing the name of the founders of wineries or their domains. The most widely known brand is Koudiat, the best of which comes from certain vineyards around Tunis. Koudiat and Rossel, another fine wine, are supple and friendly, full-bodied and with an attractive color of dark red brick. The rosés are sturdier than rosés from Provence, and often have a tinge of amber close to *pelure d'oignon* (onion skin). They are usually dry, flavorful, and with a pleasant though peculiar flavor suggestive of the scent of new-mown hay.

The wines of the domain of St. Joseph, in the district of Thibar some 60 miles west of Tunis, are today the best in Tunisia. The Jesuit fathers who work the domain (and also operate St. Joseph school) produce some 250,000 gallons of wine a year from their own vineyards, and also process harvests from the immediate neighborhood. The best St. Joseph wines are red, and should be aged. They have a solid constitution and a strong bouquet, and their deep color is reminiscent of some Châteauneuf-du-Pape wines. Both reds and rosés come from the Carignan, Alicante, Grenache, Cinsault and Mourvèdre grapes. Lighter and more refreshing than the reds, these rosés are particularily well suited to local consumption. White wines come from the Pedro Ximenes, Clairette, and Ugni grapes, and from the local Abeidi.

Sweet wine is made from Muscat of Alexandria or Grenache. The must is left to ferment to only 6 or 8 percent of alcohol; then it receives high-proof wine brandy to bring it to an alcoholic content of 15½ to 17 percent. Germany is the chief purchaser of these "Muscat de Thibar" and "Djebel Doré" wines. Other wines are made with Muscat de Carthage, from Kélibia and Cap Bon.

Tunisian red wines past the age of three or four years are entitled to be called old, and they are usually exposed to one major shortcoming: being brought to "room temperature," which in Tunisia is like receiving a steam bath. Tunisians, as a rule, possess no cellars, and during the summer an average temperature of 105 degrees is not uncommon from 10 A.M. to 5 P.M., except on the seashore.

Carthaginian vineyards, five centuries before our era, produced much wine that was drunk during festive parties. These are depicted in many ancient mosaics. But Tunisians are sober, if not totally abstinent. A few of them do drink wine, perhaps because it is less costly than the national *boukhra*, distilled from figs. But when they drink, they drink in cafes rather than at home.

The Tunisian vineyards, which covered 40,000 acres in 1910, now spread over 125,000 acres, despite the ravages of the phylloxera in 1936. In May, 1964, eight years after it gained independence, Tunisia nationalized large properties, and the agricultural activities of Europeans were interrupted, with a few exceptions. Private wineries are today in the minority. With its eleven wineries, the *Union des coopératives vinicoles de Tunisia* (U.C.T.V.) is the largest producer.

In retaliation for expropriation, the French government canceled the arrangement by which a quota of Tunisian wine was purchased in France at a price double the current market rates. But Tunisia survived the crisis, and the situation has since improved. New arrangements have been made with France.

MOROCCO

Like Algeria and Tunisia to the east, Morocco has two climates—one, inland, dry and humid, the other more moderate, being tempered by the sea and by a more varied and mountainous terrain. Wine growing has existed since antiquity, but it was only after the French established their protectorate in 1912 that a modern wine industry started to develop alongside the old traditional vineyards.

The vineyards grew rapidly. One reason was the influx of immigrants from Europe who were at least drinkers of wine if not makers of it. Despite the ravages of the phylloxera in the 1930's, the rehabilitation of Moroccan wine growing has continued at such a pace that the area of the vineyards has multiplied nearly tenfold since 1912—from about 17,500 acres to 150,000 acres today.

Today, in independent Morocco, vineyards alternate in various regions with orange, olive and other fruit trees, with vegetables of all sorts, with cactus and

Aleppo pines and with eucalyptus and oak. To the west they extend along the Chaouia and Doukkala coastal plains, and into the Rharb region further north. Eastward they lie in the valleys of the Sebou River and its tributary, the Inaouene, while southward they are found on the slopes surrounding Marrakech.

The small vineyards, where the bulk of the work is still done by hand, are gradually disappearing. They exist chiefly in the hilly regions. Large wineries, modern and well equipped, are found on the rich plains around Casablanca, Rabat, Meknès, Fez, Taza, and Oujda. In these vineyards huge tractors are taking over the work of mules and horses. Spraying is done with a battery of jets hanging on both sides of the spraying tank—and sometimes even with the help of small airplanes and helicopters.

The best wines, which are usually table varieties—and unexpectedly light for such a climate—come from around Meknès. However, aside from an agreeable Gray Wine called Boulaouane and one or two others, Moroccan wines are little known outside the country, and none can reach the excellence of a good Mascara or an aged Thibar wine.

GRAPE HARVEST AT BOULHAUT.
(*Moroccan Tourist Office*)

GREECE

Young Dionysus, son of Zeus and of Semele (the daughter of the King of Thebes) traveled one day to the island of Naxos. There, according to legend, he saw a plant so beautiful and frail that he wanted to take it home. He uprooted it gently and placed it in a bird's bone to keep it alive; but the plant grew so fast that its roots started coming out at both ends of the bone. Dionysus then found a lion's bone, into which he introduced the plant and the bird's bone. As the plant went on growing, the young god found an ass's bone and inserted the roots, together with the bird's bone and the lion's bone, into it.

Returning to Nysa, where he had grown up under the care of nymphs who protected him from Zeus' jealous wife, Hera, Dionysus found the roots hopelessly entwined with the bones and planted everything into the ground. The vine—for such it was—grew and yielded magnificent grapes, and Dionysus made wine that he gave men to drink. When they drank it, the men sang like birds, and when they drank more, they became strong as lions. But when they drank too much, their heads drooped and they became as stupid as asses.

When Dionysus grew to manhood he acquired divine status and exceptional powers, but also a touch of madness. The power he most frequently resorted to as a punishment to those who offended him was to turn them mad. But to those who welcomed and pleased him, he gave the gift of wine.

Frescoes, sculptures, and written accounts have survived to give a fair idea of the varied wines and viticultural processes of ancient Greece. Hesiod, the Greek poet of the 8th century B.C., mentions some viticultural techniques: the vine shoots were pruned in early spring, trenching was completed in May, weeds were uprooted from the vineyards, and clods of earth broken up. The harvesters were gay, lightly clad youths flocking to the country at harvest time. Baskets full of grapes were carried to the pressing area, usually outdoors, sometimes in a cellar. This went on to the rhythm of the grape-pickers' songs or to the accompaniment

BACCHUS AT REST. Greek. *Louvre. (Photo Giraudon)*

BACCHANTES. Bowl from Corinth, 6th century B.C. *Louvre. (Photo Chapman)*

of Dionysus' beloved flute. Grape-crushing became a dance, which has remained to this day in Greek folklore.

Wine was collected in large jars lined with plaster or pitch, and some of it was boiled down. A wine boiled down to half of its original volume probably corresponded to the aperitif-type wine of our days. Jars were closed with pitched cork or kept away from oxygen by a layer of oil and stored in cellars.

Resin was used as a preservative, and it is still used today for the popular Retsina wine. An amazing number of substances were used to flavor the wine: infusions of flowers of the vine, leaves of pine or cypress, bruised myrtle berries, shavings of cedarwood or southernwood, bitter almond, spikenard herb, saffron, clover, or sweet scented flag. For clearing or preserving wines, the ancient Greeks used pitch, chalk, sand, pounded seashells, toasted salt, ashes of wine stalks, roasted gall nuts, cedar cones, burnt acorns, olive kernels, or sweet almonds. Or else a lighted torch or a piece of red-hot iron was plunged into the wine.

Greek wines of today are typical of a warm subtropical climate in that they are often high in alcohol, sweet, and lacking in finesse. Nevertheless many are fair, and taste particularly good when drunk with the clear bright sunshine of this hospitable country. The most current wine is white Retsina—usually served right out of the barrel, available anywhere, and less expensive than mineral water. Made chiefly with the Savatiano grape, Retsina contains pine-tree resin and has a very, very special taste of its own. "You like it—or don't" is the most frequent judgment upon Retsina wine, but this is not quite so: one grows to like Retsina. It is particularly pleasant in summer, and helps one become accustomed to the Greek cuisine, which is often greasy. Piping hot *moussaka* (a meat pie with eggplant and mashed potatoes, topped with cheese), *dolmades* (packets of rice wrapped in vine leaves marinated in salt), and *keftedes* (meat balls) seem to clamor for Retsina, and a snack of *micropsaria* (assorted small fish) is pleasant with Retsina.

If you wish to taste and compare many Greek wines one after the other, you should be in Athens during the month of September. An annual wine festival is held in Daphni, a few miles west of the city, next to the famed Daphni monastery, one of the most beautiful Byzantine monuments in Greece. The wine festival takes place in an enclosure, which can be entered for a modest fee that entitles you to a glass and decanter. After that you can enjoy the free samples of wine as you wander around the stands.

Some of the best table wines come from the Peloponnesus and from the islands: the brownish-red Kokinelli (not a brand, but a type, which differs from region to region and island to island); the Cretan factory-made King Minos, potent but often good; and the Rhodian Chevalier de Rhodes, probably the most consistently good bottled red table wine. Paros, Lemnos, Patmos, Naxos, Santorin, and Corfu have fair red wines, but the wines bottled by the factories at Patras and in Attica, under the label "demestica" or "kampa," either red or white, are usually ordinary and uninteresting. Santorini, the volcano island that some scientists have identified with Atlantis, has an excellent dry white wine, sold in half-gallon, wicker-covered bottles. Macedonian Kozani red wine is heavy but quite good, and the sweet, natural Siatista, made from the black Muscat, becomes good after eight to ten years of ageing.

The island of Samos has a highly organized wine industry, and has been replanted since the phylloxera chiefly with a white grape, the French Muscat de Frontignan. Wine is made by adding alcohol to the juice during fermentation: the result is a strong, sweet wine, with a pronounced muscat flavor. Rhodes also has some Muscat, and the Malvasia grape is found nearly everywhere. The name originated from Monemvasia, the Byzantine and Frankish city perched on the rocks at the tip of one of the Peloponnesian fingers; but the best Malvasia (or Malmsey) wines today do not come from Greece but from Sicily. An excellent dessert wine is made in various districts from the Mavrodaphné grape (known in the Balkans as Mavroud); some like it as an aperitif.

Viticulture is important to Greece—it provides a livelihood for more than a tenth of the population, and its products—wines, raisins and grapes—represent a quarter of all Greek exports.

Cyprus is a large producer of wines, many of them fortified and sweet. Wine growing was given a strong impetus by the Crusaders, who did not produce much wine in the embattled Holy Land, but shipped it from Cyprus. Richard the Lion-Hearted celebrated his marriage to Queen Berengaria in 1191 with Cyprus wine, and the Knights of the Order of the Templars and of the Order of St. John of Jerusalem made the golden Commanderia wine, which was—and still is—exported to Western Europe. Another excellent wine is the Muscat of Omodhos, reminiscent of Samos wine.

IN GREECE, the family wine store consists of a small barrel. *(Photo Almasy)*

THE CANAAN GRAPES, detail from Poussin's "The Promised Land." *Louvre. (Photo Mercurio)*

ISRAEL

The Bible, as we have seen, has countless references to wine. It was the prophet Isaiah who ordered Moses to send scouts to the land of Canaan, from where they brought back a vine stalk and a bunch of grapes. Christ's first miracle, the changing of water into wine, took place in Cana, now Kfar Cana, a blossoming little Galilean village. Wine was plentiful also during the Roman occupation and in the first centuries of Christianity, until the first Moslem conquest of Palestine. In 996, however, the Caliph Al-Hakim ordered the destruction of all vineyards whose grapes could be used to make wine.

There followed a long period of viticultural drought—briefly interrupted when the first Crusaders clattered into the Holy Land. They found abandoned vineyards but also a few cultivated ones, in the more isolated, forgotten Christian communities. A Crusader called Burchard wrote a diary which mentions Basak, a village near Bethlehem, "which makes wine without equal . . . As the Moslems don't drink wine, they have uprooted all the vines except those growing around Christian communities, where it is sold at great profit." There was not much of it, and consequently the Crusaders imported their wines from Cyprus.

In 1882, a few Jewish settlers pitched their tents or built their shacks in Turkish Palestine, on a barren hill of the Judean plain a dozen miles southeast of

Jaffa, and proudly called their village Rishon-le-Zion—the "first in Zion." They planted vines, but water was scarce, the soil was poor, and they despaired at the meagerness of their crop. Then from France, the land of wine, came a man who envisaged a great future for Palestine's wine-growing industry. He was Baron Edmond de Rothschild, owner of vineyards in the region of Bordeaux. He imported the finest European plants and sent experts to teach the settlers modern methods of wine growing. In 1889 Rothschild financed the building of a modern winery at Rishon-le-Zion, and huge cellars which rivaled the best in the world. A few years later he built another winery at Zichron Jacob, southeast of Haifa in the hills of Samaria. It was the birth of the *Société coopérative vigneronne des Grandes Caves,* also known as the Carmel Winegrowers' Cooperative, the largest in Israel today.

Fighting between the Turks and the Allies during World War I and an invasion of locusts practically wiped out the vineyards, and production remained low until 1948, when Israel became a state (and one which can claim, rightfully, to be the world's youngest, as well as the oldest, wine-producing country). By 1965 the area of the vineyards exceeded 25,000 acres in seven principal wine districts, including man-made oases in the Negev desert. There are some twenty wineries, most of them Kibbutz-style cooperative farms.

All the wine made in Israel is, of course, kosher (Hebrew, *kashrut*), and as such it is a commodity much in demand by Jews the world over. Kosher rules for wine making are described in detail in the Talmud, the body of Jewish civil and canonical law, and some of these rules make good, plain oenological sense. *Orla,* for instance, forbids the Jews to drink the juice of the grape during the

TERRACED VINEYARDS on the hills of Judea. *(Photo Werner-Braun)*

ORNAMENTAL SCULPTED FRIEZE of grapevines.
Coptic bas-relief, Egypt. *British Museum. (Photo Boudot-Lamotte)*

first three years after the planting. In fact, for most varieties of grapes, no wine would be made until the fourth year anyway. On the fourth year the fruit of the vine should be brought to the Temple in Jerusalem, and from then on, the vineyard is *kashrut* for wine making. The Chief Rabbinate in modern Jerusalem has a department to enforce *Orla,* and it keeps lists of new vineyards perhaps more accurately than the Department of Agriculture. Before the harvest, a rabbi usually inspects the vineyard.

Another rule forbids mixed culture: no vegetable can be planted in the vineyards. And another, which must have made oenological sense two thousand years ago, though less so today, bans food and bread and anything that ferments from the place where wine is made. When entering the winery, the worker leaves his lunch-bag outside.

Some of the rules, however, are of a strictly religious nature: a small part of the produce of a winery must be given to the Cohens, the hereditary priests of the Temple. The Cohens are many today, but they do not line up at the winery when the wine is ready. Instead, the custom has been accepted of pouring 1 percent of the harvest (or a symbolic amount thereof) onto the ground. *Shomer Shabat* requires that every worker in the fields and in the winery, even if not a Jew, observe the Sabbath—which, in Israel, is the official day off anyway. Another rule is that a non-Jew should not touch a bottle of Passover wine.

Many Jews turn to Israeli exports not only because the wine is *kashrut* but because there is a greater variety. And Israeli wine is likely to be better than a non-Israeli kosher wine—notably better than one made in the United States, where kosher wines have become identified with undistinguished varieties characterized by a cloying sweetness, although sweetness is not specified in the Talmud. Israeli exports of wine today reach some forty countries, and growers now hope that the local consumption—barely over a gallon a year per capita—will increase.

No district names are used to label Israeli wines because nearly all are blends. Most have brand names, such as Carmel Oriental, Carmel Avdat, or Old Yakenet Wine. Labels no longer carry European place names such as Champagne, Burgundy, or Sherry, since in 1966 Israel signed the Madrid Pact, which limits certain geographic names to wines produced in those places. There are white, rosé and red wines—some of the latter reminiscent of Clarets, others of Burgundies, though few have claim to any distinction. The grape varieties were selected for a generally hot, arid climate: for red wines, Alicante, Alicante-Bouschet, Carignan, Grenache, some Cabernet Sauvignon, and even American *labrusca* plants, such as the slipskin Concord; for white wines, Muscat of Alexandria and of Frontignan, Clairette, the excellent French Sémillon, and the Ugni Blanc.

In 1929, it could be said of wines from the Holy Land that none were good and only a few were passable. But since then quality has improved. From 1953 onward, many Israeli wines have won prizes at international wine fairs—for example, Bourgogne Château Rishon and Bourgogne Château Windsor ("Bourgogne" has since vanished from the labels). Dessert wines such as the Port-type Partom and Sharir are appreciated by the Israelis. Edom Atic dry red wine, much of which is exported, is honest and reliable, and the pale pink Rosé of Carmel is honorable. Avdat—named after the biblical town—is dry, white, and a passable table wine. Some wine lovers hold that too many of Israel's wines are still oversweet—but then, many who buy them, notably in the United States, seem to like them that way.

What used to be Israeli "Champagne" has taken a new name: President. The makers do not even mention that the Champagne method is used—although it is. Every bottle merely states, that the wine is "fermented in this bottle by the traditional method." The President *brut* of a good vintage year can be excellent, and the sweeter rosé is pleasant.

LEBANON, SYRIA, JORDAN

Warm and sunny weather prevails in Lebanon from May to November, streams flow gaily, cedars and poplars murmur in the breeze, and a clement sky smiles on the winegrower. Phoenicians grew wine, and church vineyards near Tripoli were planted during the Crusades. The production today is not quite a million gallons a year. It comes chiefly from plants imported from the Languedoc in France: Cinsault, Carignan, Aramon for the reds, and Ugni, Clairette and Téramon for the whites. A sweet Muscat-type wine is made with Grenache and Moscatel. The Karm El-Mir plant gives a celebrated "gold and sweet wine," which the Lebanese occasionally like to drink with Oriental pastries filled with almonds or pistachios and soaked in honey or in sugar syrup perfumed with almonds. Château Masar near Beirut produces a fair red wine—suited to accompany Lebanese specialties such as *lahem mechoui* (mutton on the spit) or *chawarma*, thinly sliced mutton layered with fat to form a sort of pine cone. The cone is flavored with spices, lemon juice, and mint leaves, and cooked while it revolves before a fire. The trend in recent years has been toward rosé wines, which are dry and a bit acid, fruity and with an alcohol content of about 12 percent.

In Syria and in Jordan, wine is made chiefly in convents and abbeys. The strong and plain reds go well with the local *kebabs* (skewered meat) or *kebbes* (meat cooked with wheat).

None of the wines from this region are exported—save for a few bottles shipped to homesick Lebanese businessmen in England or in Africa.

EGYPT

The oldest known wine list has been unearthed in the pyramid of Pepi II, fourth king of the Sixth Dynasty, who reigned in Memphis some 40 centuries ago. It mentioned Delta wine, white wine, Pelusia wine, red wine, and Latopolis wine. Other archeological findings include amphorae in which there were undisputable traces of wine and honey. And references to wine in hieroglyphics are countless.

Is there any reason why modern Egypt could not produce the wines it was famed for several thousands years ago? This was the question that an enterprising Greek, Nestor Gianaclis, owner of an Egyptian cigarette factory, asked himself in the 1880's. To find out, Gianaclis purchased a few acres of land not far from El Alamein and planted them with vineyards. The vineyards grew. Eighty years later, in the 1960's, vineyards in Egypt covered more than 20,000 acres, and Egypt's annual yield was 750,000 gallons. This is far below what it used to be, and, as a rule, the quality leaves much to be desired.

The vineyards have so many varied plants that they are like viticultural museums. Besides Chasselas, French Pinot and Gamay grapes, and Italian plants, there are numerous little-known native growths, such as Fayoumi, Guizazi, white and red Roumi, and many sumptuous table grapes, protected from the *khamsin* (sand wind) by eucalyptus and the local varieties of cypress.

Watering is the chief problem. If the Nile is low, the *fellah* has to carry on a ceaseless, tiring work that hasn't changed since the time of the Pharaohs: this task, the *natala,* consists of filling up a water basket in a canal, and pouring it out into irrigation gutters. Another irrigation system is the *sakia,* a wheel with cups, turned by ox- or donkey-power. The harvest in this sun-drenched land is a picturesque sight. Men armed with fine-toothed pruning knives clip off the heavy grapes and throw them into wicker trays, which women balance on their heads and carry to the winery. If grapes have to be taken a long distance, they are placed in baskets made of palm tree leaves, which are piled on chariots or in trucks (the former being far more reliable than the latter).

These efforts enable the visitor, dining in one of the typical rooftop or boat restaurants in Cairo, to have a Clos de Matamir with his roast pigeon, or a Clos Mariout with lamb kebab, or a rosé while he watches the belly dancers. The giant shrimp *(gambari)* or the Egyptian caviar *(batarakh)* call for a Ptolemean growth, which is white, pleasant and refreshing.

TURKEY

Wine was abundant in Thrace, Anatolia, and along the Aegean coast throughout antiquity, but the Moslem religion banned its making in Turkey and in the huge empire conquered by the Turks from the 14th century on. Wine making became legal again after the founding of the Turkish Republic by Kemal Atatürk in 1923, and production has been steadily climbing ever since. A wine research center was founded in 1929 to modernize archaic methods and put some order into the viticultural chaos resulting from the indiscriminate growing of some 200 to 300 local grape varieties. Many fine plants were imported—Sauvignon, Sémillon, Pinot Blanc and Noir, Traminer, Cabernet and Merlot—and the annual production reached six million gallons, about half from large state-owned wineries and half from smaller, private ones. There is a lively export business, chiefly to Germany and the Scandinavian countries.

Most of the wines are *beyez* (white), a few are *kirmizi* (red) or *pembe* (rosé). The label should be carefully examined. If it says *sarap*—sweet—it's probably too much so, just the way the Turks like it.

Guzel Marmara is a dry white wine produced in Thrace from the Yapinçak grape, one of the most widespread native varieties. Straw-colored with a tinge of green, it has a slightly bitter taste of fruit and seeds which makes it quite refreshing. Another dry white wine, Izmir, is made on the Aegean coast from the Sultaniyeh, a seedless grape. It has a golden-yellow color, a very slight aroma, and contains 11 to 12 percent alcohol.

Among red wines, there is the Narbag, from the Narince grape. It has a slight fruity aroma, but is rather too sweet. Kalebag red wine, made near Ankara with

HITTITE WINE JARS, found during excavations near Antioch. About 2,000 B.C.
(*Photo Turkish Government*)

the Kalecik black grape, has up to 13.5 percent alcohol, a light aroma, and much snoothness coming from a high glycerin content. Buzbag is made with a grape known as Oküs Gözü ("bull's eye"); it is dark red with bluish reflections, and ages well. So does Bögazkar—"the strangler," so named apparently because its high tannin content may grasp the drinker at the throat. Traka wine, from Thrace, made from a vine called Papaz Karasi, is dark red and full-bodied, and is usually bottled in its third year.

IRAN

A Book of Verses underneath the Bough
A Jug of Wine, a Loaf of Bread—and Thou
Beside me singing in the Wilderness—
And Wilderness were Paradise enow.

The reader of "The Rubáiyát" of Omar Khayyám (the Edward Fitzgerald version) may expect to see a small vaporous genie escape from a long-necked, straw-bound flask of pungent and spicy Persian wine. And perhaps one will—if the reader can secure such a rare bottle.

Persians, like many others, claim to have discovered wine. Their story is that many years ago, there lived a great Shah of Persia, called Shah Djemsheed. He loved fresh grapes, which were regularly brought to him by Gulnare the Fair, his favorite concubine. But one day she forgot to bring fresh grapes, and the Shah found spoiled ones at his bedside. Desperate, Gulnare ate them, hoping to die of poisoning. But she only slept, and had such amazing dreams that the Shah also tried the grapes. Both thereupon became slaves to "the delicious poison."

The best-known Persian wine today is Shiraz, mentioned by Marco Polo and said to have existed at the time of Persepolis. The vine stalks can be as thick as tree trunks, and the wine making methods are ancient. The barefoot-trodden grapes ferment for three weeks, and the wine is stored in earthenware jars of more than thirty gallons each, called *guarabars* (or carboys). The Shiraz grape may be related to the Syrah grape used to make Hermitage red wine, but not to the Petite Syrah, widely used in California. A few other, rare wines are produced near Teheran, Ispahan, and Tabriz—but the total production seldom exceeds 50,000 gallons a year.

COPTIC FRIEZE with grape motif. *Louvre. (Photo Chapman)*

Eastern Europe

SOVIET UNION
HUNGARY
BULGARIA
RUMANIA
CZECHOSLOVAKIA
YUGOSLAVIA

HUNGARIAN WINE DRINKERS. Porcelain gourd.
Wine Museum, Tokay. (Photo Chapman)

Preceding page: **THE TOKAY WINE MARKET,** Hungary. *(Photo Chapman)*

THE SOVIET UNION

A wine lover who recently visited Moscow remembers a distressing episode which took place in a fashionable restaurant off Gorki Street. He had been given a seat at a table occupied by four local youths who were ordering dinner. Before looking at the menu, the Russians called the waiter and ordered: 1) a bottle of Armenian cognac; 2) a bottle of vodka; 3) two bottles of Georgian white wine.

They drank the cognac first, while waiting for their food. Then, as the meal was served, they washed it down with small glasses of vodka, each followed by a glass of white wine. Fascinated, the wine lover asked his neighbors to explain this sequence. "It's simple," answered one of the Soviet diners: "You drink the cognac first, because it's the tastiest of the lot. Then you take the vodka, because it's a Russian tradition. And you drink wine to quench your thirst. Besides, cool wine feels good after the vodka's burn."

As the evening went on, the wine lover further learned that in the Soviet Union the production and drinking of wine is encouraged to combat alcoholism by fostering wine as a substitute for vodka, the "little water" which is absorbed in the Soviet Union in literally staggering amounts. The plan to expand wine production, and to raise vodka prices to a drastic level, was adopted during the 21st Congress of the Communist Party in 1956.

In the decade following this decision, the production and drinking of wine have nearly trebled. To the distress of party planners, however, this has not reduced the consumption of vodka.

Nevertheless, much has been accomplished in Soviet viticulture and viniculture, chiefly by putting modern science and technology to work. Soviet oenologists have to their credit a number of genuine findings and accomplishments, some of them useful, others at least entertaining. Biochemists, for instance, have calculated the size of Champagne bubbles, and have found a way to produce *champanskoye* in three weeks. Botanists have developed, through selection, foot-thick vine stalks to survive below-zero temperatures in Siberia. Physicists are using ultrasound to speed up ageing of such local wines as "Madera" or "Chato Ykem." And chemists at the Moldavian Research Institute for the Alimentary Industry have even threatened to replace the traditional human wine taster with an electronic device that can be "fed" with the olfactory and gustatory memories of a wine lover, and pass judgment on a wine.

Wine in the countries that make up the Soviet Union has a long history. Wild, edible grapes existed long before man—notably the *Vitis silvestris,* a *vinifera* subspecies, which can be still found in Europe and Asia. Soviet botanists have found at least sixty varieties of wild vine on the shores of the Black Sea alone, and some of these are believed to have been domesticated by Stone Age man. Wine making was widespread during the Bronze Age, at least in Georgia, where several artifacts of that time have been unearthed bearing the images of men feasting with wine. Homer praised the "sparkling, foaming wine of Colchis" (western Georgia today).

The earliest known state formed on the European territory of today's Soviet

IN A RUSSIAN TAVERN. Popular print.
Bibliothèque nationale, Paris. (Photo Etienne Hubert)

Union was the short-lived kingdom of Urartu—or Van, from the lake of that name on the Armenian plateau—which held considerable power from the 9th to the 7th centuries B.C. In Urartu, water from mountain streams was channeled to irrigate the vineyards in the valleys. Huge cellars have been found, dating back to the 8th century B.C. And when Assyrian King Sargon invaded Urartu in 714 B.C., his soldiers looted the wine cellars; bas-reliefs show Assyrian warriors guzzling the wine through tubes.

In the kingdom of Bactria (today the Uzbek Soviet Socialist Republic), invaded by Alexander the Great in the 4th century B.C., the vine was also cultivated, along with wheat and rice. Roman historian Quintus Curtius Rufus wrote that Alexander found many beautifully irrigated vineyards, which were further developed during the Macedonian occupation.

The Romans who settled on the sunny Crimean shores at the beginning of our era introduced Roman plants. A few Christian monasteries were later founded, producing altar wine, but these fertile southern lands were desolated by successive invasions of Mongols from the east and Turks from the south.

By the 15th century nearly all of the wine-growing areas were under Moslem domination. Vinicultural traditions crumbled under the influence of this religion, which forbids alcoholic beverages. They survived only in a few monasteries, and in what is today the Moldavian S.S.R., northwest of the Black Sea. Here organized trade fairs attracted many Western European wines that eventually reached the Russian Court. It was also in the 15th century, however, that vodka, sold in government owned *kabaks* (taverns), contributed to destroying the tradition of wine. Yarilo, the Russian counterpart of Dionysus, reigned supreme—not as the God of Wine but as the God of Vodka. On May 30, the last day of the annual Yarilo festivities, revelers throughout Russian towns and villages could be seen sprawled on the streets, dead drunk.

The capture of Astrakhan on the Volga by the Russians in 1555 marked the beginning of the withdrawal of the Moslems from the south, and a of renaissance of wine growing which started spreading, with Christianity, from the surviving monasteries. Much war booty came to the Russian nobility in the form of Crimean land, where they made wine and shipped it to Moscow or St. Petersburg. At the Kremlin, etiquette prescribed the drinking of white wine from silver goblets, red wine from gold ones.

At the beginning of our century, Russian vineyards were decimated by phylloxera, and had only started recovering when World War I and civil war broke out. In the first years after the Revolution, little was done to replant vineyards.

Today, however, wine flows abundantly from wineries situated mainly in the southern Soviet Union but also in Siberia and in the Eastern Asian republics. Most wines are produced by *sovkhozes*, "avant-garde" agricultural establishments where farmers are on salary, and from *kolkhozes,* where salaries are lower but are compensated for by small "personal" lots of land.

HARVESTING GRAPES IN MASSANDRA,
near Yalta in the Crimea. *(Photo Ginette Laborde)*

Four hundred native vine stocks of eating and wine-making grapes are grown in the Soviet Union. Western plants—Muscat, Riesling, or Cabernet—grow side by side with native plants such as the Baian-Chire of Azerbaidjan, the Rkatseli and Saperavi of Georgia, the Blue of Altai, or the Emerald of Kuibyshev. Many European-American hybrids are also used, notably in Moldavia.

Most Soviet wines, however, do not flatter Western palates: the Soviets have a huge sweet tooth.

NOTABLE SOVIET WINES

Several Soviet wines deserve a mention: a red wine called Tch-Haveli No. 7 can be quite good, but it is not easily found. Likewise, the Georgian Hvanch-Kara (which was Stalin's favorite) is scarce. Mukuzani No. 4 is the best-known red wine, and has some of the qualities and fruitiness of Beaujolais. There is a red **Бордо**; it is pronounced "Bordeaux," and the resemblance ends there. The factory or farm where the wine has been made, or the name of the wine-making firm, is indicated on the bottle; there is also a number (Mukuzani No. 3, No. 4, etc.) which corresponds in principle to a blending formula developed in oenological laboratories or institutes. Thus, among relatively interesting Georgian white wines, there is Tsinandali No. 1, dry, fruity and the most current dry table wine in the Soviet Union. Saperavi No. 5, Teliani No. 2, and Tibaani No. 12 waver between almost dry to sweet, while Gurdzhaani No. 3 is dry, though often devoid of any personality. Among red wines, Guinz-Malauri No. 22 or 26 and Akhasneni are acceptable.

There are many sweet dessert-type wines, which are often drunk as table

Left: WINE TASTING in
the Urjaani region
of Georgia, Russia.
(Photo A.P.N.)

Opposite left: THE OFFICIAL WINE
TASTER of Abrau-Dyurso,
on the Sea of Azov, Russia.
(Photo Paul Popper)

Opposite right:
RUSSIAN "CHAMPAGNE" is
circulated in these vats
to assure quick maturing.
(Photo E. Kassin)

wines; the Muscat Massandra or Magaratch can be excellent, and the Georgian Salkhino No. 17 is fine. It takes a special knack to figure out some of the Soviet wine labels, which often try to strike a chord in the prospective buyer's heart rather than to be informative. There are, for instance, wines called Pearl of the Steppes, Sunny Horizons, and Dark Eyes. Sometimes, however, there is a direct reference to the type of plant used, such as Transcarpathian Riesling or Preskovieski Muscat. The alcoholic content is usually indicated—if at all—with casual approximation: "8 to 10 percent" or "16 to 19 percent." Bottles bearing the same label can contain wines from different origins, bottled in different places. Samtrust, for instance, the largest Georgian wine trust, which produces forty out of the sixty varieties of wines in the Georgian republic, has several branches, and each makes its own blends. Yet each of the branches will produce, say, Tsinandali No. 1, a number theoretically representing a blending formula, but the wine obviously varies, since its ingredients come from different Georgian districts.

Most wineries have an industrial aspect. Blending and racking are directed from a central control post, and take place through miles of glass pipes attached to the ceilings of the cellars. Most of the wines are stored in glass-lined cement vats, but the ancient earthenware *karass* is still used.

The most beautiful wine-growing region in the Soviet Union is Crimea—the Soviets' favorite summer resort. Around the cliffs the road runs between palm trees and orchards, pine and oak groves, palaces built by the tsars, and palatial *dachas* by modern rulers. Grapes grow all over the peninsula, and some of the locally made wines are quite good. One of the best is the Massandra, a sweet wine that has been produced in a seashore château for over a hundred years.

GEORGIAN WINE LABELS. Left to right: Tsinandali, Tvickin, Salkhino, Kardanakhi, Naareouli, and Khvantchkara, semi-sweet "Champagne." *(Photo A.P.N.)*

Georgia, too, has excellent *vins du pays,* not known outside the republic. Georgians try to avoid drinking factory wines, and in remote Georgian villages, where ancient feudalism is not quite dead and where Georgian "princes" are still treated with respect by erstwhile serfs, a few fortunate travelers have been introduced to rare vintages. To these growths Georgians ascribe their vigor and longevity, noting that there are more centenarians in Georgia than anywhere else.

CHAMPANSKOYE

A matter of particular interest in the Soviet Union is "Champagne." Traditionally the wine of fair ladies and dandies, it was sung by poets in Russia as much as anywhere else. Wrote Pushkin:

> *The pail is brought, the ice is clinking*
> *Round old Moët or Veuve-Cliquot.*
> *This is what poets should be drinking*
> *And they delight to see it flow.*

But now *Champanskoye* belongs to the comrades. In Moscow the harried factory worker can stop off at a counter on Gorki street or elsewhere, stand in line, plunk down sixty-five *kopecs* (70 cents), for which he receives a ticket, and stand in line again to exchange the ticket for a three-ounce glass of *Sovietskoye Champanskoye.* The wine is likely to be manufactured in the Moscow Champagne Factory with a new, super-automated process, which has recently been patented in the United States and in several European countries. (Patent 3,062,656, issued in November, 1962, in Washington, was the second to go to a Soviet citizen in a decade.)

The Soviet method of making Champagne may have removed some of the poetic attributes from the wine but it has introduced into it adequate amounts of sugar. A number of Soviet scientists have studied bubbles in sparkling wine, "which occur when the equilibrium of CO_2 in the oversaturated system which is Champagne is disrupted." According to reports published by Professor Gregori Guerassimov Agabaliantz of the Krasnodar Alimentary Institute, the physicochemical phenomenon known as a bubble first arises when the bottle is opened. The bubbles form around "cavitation nuclei" and rise, one after the other, until

the gaseous saturation of the wine becomes too low. A very important thing about bubbles, he notes, concerns their surfaces: the bubbles, in order to be long-lasting, should be wrapped in large organic molecules, "tensio-active macromolecules." Hence the superior quality of natural over artificial bubbles, for the natural Champagne-making method favors the autolysis—or self-digestion—of the yeast, which leads to the formation of such macromolecules.

Having thus dissected the bubble, Professor Agabaliantz and his colleagues concluded that in order to make wine with a good sparkle and foam, it was desirable to develop a technique encouraging the formation of large quantities of small bubbles wrapped up in those large macromolecules. Soviet oenologists went on to develop a new method known as the "continuous stream technique," in which the second fermentation of wine (which makes the bubbles) occurs in vats instead of bottles at a constant pressure about five times that of normal atmosphere at sea level. And as in the bottle method, the process takes place in the absence of oxygen.

"With our method," says Sergei Alexeievitch Brussilovsky, one of the inventors of the "continuous stream" and now the debonair and efficient director of the pioneering Moscow Champagne Factory, "we obtain wine which many experts cannot distinguish from Champagne made in the old-fashioned way." *Sovietskoye Champanskoye,* he believes, could be as good as any Champagne provided the same good blend of wines were used, instead of the wine received by the Moscow factory, which comes mainly from Riesling grapes. These have a high yield but less *finesse* than the Pinot Noir, Pinot Meunier or Chardonnay used in France.

The method itself is relatively simple. The wine blend, shipped by truck and train to Moscow from the south and kept in cement, glass-covered vats, is first introduced into two 1,300-gallon tanks, in which the second fermentation process will be initiated. Yeast is grown separately, and pumped into the blend to be fermented. The yeast is introduced into the wine *after* culture, when it no longer needs to multiply. This allows the fermenting wine to be kept under constant pressure, in the absence of oxygen, and at a temperature which slowly decreases from 68 to 32 degrees Fahrenheit, conditions that do not allow the yeast to multiply.

The wine, still under five atmospheres of pressure, is pumped into the first of the fermentation vats. And it goes on flowing in a continuous stream through six more vats, where it accumulates the required amount of carbon dioxide, amino acids (the famous macromolecules), vitamins, and other natural yeast byproducts. By the time the blend leaves the fermentation vats, it has been cooled off to approximately water-freezing temperature, which favors the precipitation of deposits. The surface of the liquid in the receiving and discharging tanks is covered with a thin layer of deodorized vaseline or paraffin oil to reduce the absorption of oxygen. Champagne liqueur, containing some 70 percent sugar, is added, and the mixture enters a rapid heat-exchange vat where it is rapidly cooled to about 23 degrees "to increase the stability and quality of the wine."

The Moscow factory—a two-storied installation housed in a nondescript red brick building on Ossipienko Street, not far from the Kremlin—is now entirely automated. Only a few white-coated employees are required to control the system that keeps a constant check on pressure, temperature, speed of the flow, fermentation, and other no longer mysterious events. There is a master switch crammed with electronic apparatus, glittering red and green lights, and buzzers and bells to sound the alert if anything goes wrong with the steam pressure, compressed air, or temperature.

In *champanskoye* as in other wines, sweetness is in demand. Standard varieties are called half dry, with a sugar content of 5½ percent; half sweet, 8 percent; and sweet, 10 percent. But dry sparkling wine is becoming increasingly popular.

In Russia, one drinks "champagne" *do dna*, bottoms up. The standard toasts are *na zdorovie*, "to your health," or *mir miru*, "peace to the world." If the foreigner raises the first glass and utters the second toast, the comrades will be delighted.

HUNGARY

The wine grapes throughout the rest of the Northern Hemisphere have been gathered and pressed and much of the wine has already fermented, before feverish activity becomes apparent on the slopes of a round, tall (1,115-foot) hill along the northeastern frontier of Hungary. As the sun rises on the crisp, chilly morning of October 28, St. Simon's Day, colorful scarves bob up and down among the reddish vine leaves on the gentle slopes reflected in the quiet, meandering silver of the Bodrog River below.

It's harvest-time in the Tokaj-Hegyaljai district. The slow, methodical grape gathering may go on to the end of November or even into December, sometimes under the first winter snow. The result will be Tokay wine, a truly unique essence and inimitable, though imitations have often been attempted. Rare and in great demand by the fortunate few who have tasted it, it is a wine so precious that Swiss alchemist and physician Philippus Paracelsus once traveled to Hungary hoping to extract gold from the famed Tokay grapes.

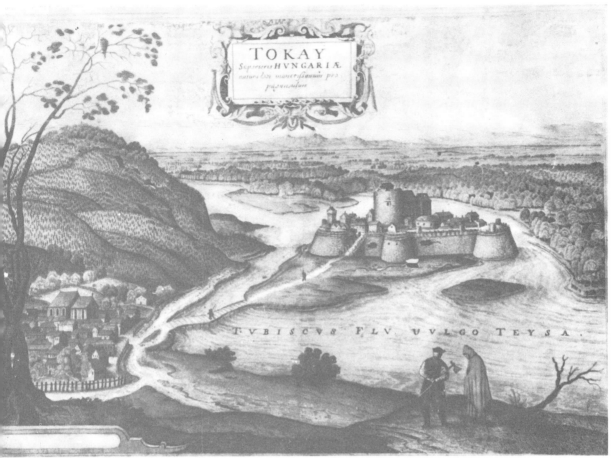

TOKAY AND ITS FORTRESS. Engraving, 18th century. *(Photo Chapman)*

MAKING TOKAY

Twenty-five small villages are entitled to the Tokaj-Hegyaljai appellation, which takes its name from the Tokaj hill, the first hill of the Eperjes-Tokaj range, and Hegyaljai, Hungarian for "foot-of-the-hill district." Wine has been made here for at least a thousand years, but at first it was considered as just another fair wine. The story goes that Tokay wines emerged from obscurity in the fall of 1650, when a war threatened the region and one Máte Lackó Sepsi, steward at the household of a Hungarian gentlewoman, had the harvest on one of the hills put off until November. The growers sadly saw their grapes shrivel and start to rot. They did not know that the mold covering the grapes was a noble one, caused by *Botrytis cinerea*, the same micro-organism that shrivels the grapes that yield the sweet and liquorous Sauternes and German *Auslese* wines. These shriveled berries, called *aszú*, are selected and picked off the bunch into the traditional *puttonyos*, six-gallon wooden buckets. *Aszú* grapes are the very soul of Tokay wines, of which there are three distinct varieties.

The best, rarest and most expensive Tokay is the *eszencia*, the essence of the *aszú* grapes, obtained simply by letting the weight of the grapes themselves ex-

AT THE TOKAY
WINE MARKET.
(Photos Chapman).

tract a golden liquid, seldom more than two or three pints from a six-gallon *puttony*. This essence holds an amazing amount of sugar—from 40 to 60 percent —and its fermentation lasts not weeks or months but years before it achieves an alcoholic content that seldom exceeds 7 or 8 percent. It is almost not a wine but a low-alcohol cordial with a fragrant *bouquet*. Little is ever drunk, since it is used chiefly to fortify other *aszú* wines.

The actual Tokay wine, or Tokay *aszú*, comes from the shrunken berries traditionally treaded by foot to avoid crushing the pips. They are treaded for many hours, until a sort of dough is achieved—so uniform that if a handful is squeezed, all of its seeps out between the fingers save the pips themselves. (Recently, a mechanical pulping machine has been designed, which is slowly making the foot obsolete.) Tokay *aszú* dough cannot be obtained every year, for such a homogeneous pulp can be achieved only when the grapes' skins are so affected by the noble mold that they melt in with the flesh.

Meanwhile, the remainder of the grapes from the district, those not affected by the mold, are pressed. They are chiefly of the local Furmint variety, also Harslevelu and Muscatel. They grow on a soil of crumbled lava on top of volcanic rocks and loess deposit, sheltered by the Carpathian hills in the north, west, and east, and open on the south to the warm summer winds of the Great Hungarian Plain. The grape juice, or must, is collected in barrels called *gönc* barrels, each containing 30 to 35 gallons. The quality of the final product de-

SAMPLING WINE from the barrel.
Note the curious pipette.
(Photo Chapman)

pends on how many *puttonyos* of *aszú* dough are added to each barrel of must. Thus Tokay wines are classified as *aszú* three *puttonyos,* *aszú* four *puttonyos,* etc. Usually, the more *puttonyos* the better the quality, though in exceptional years an *aszú* three or four can be better than an *aszú* five or six of another year.

The mixing of *aszú* dough and wine takes place in large tubs. Once fermentation begins, the mixture is stirred, poured into special bags, and treaded once again, yielding a liquid called "*aszú* must". This liquid is once more poured into *gönc* barrels and stored away.

Once bottled, Tokay wine can keep for centuries. Before World War II, the Fukier Company in Warsaw kept in its cellars 328 bottles of Tokay *aszú* of the 1606 vintage, and sizeable stocks of 100- to 300-year-old wine. These wines are said to have developed delicate aromas reminiscent of strawberry, cocoa or vanilla.

Yet another, and more common, variety of Tokay is *Szamorodni* (a Slavic word meaning something like "as born of nature") made from a mixture of *aszú* shrunken grapes and grapes untouched by the mold. Tokay *Szamorodni* can be either dry or sweet, depending on the year and the proportion of *aszú* berries, but it is always of high alcoholic content, usually over 13 percent.

Tokay wines have always been particularly favored by the Poles, and for centuries Tokay was almost a necessity of life for the Polish elite. A Polish proverb has it that "Poles and Magyars are brothers, with the sword as well as the cup." Tokay wine was also a favorite of King Louis XIV of France, who liked to call it "the wine of kings, and the king of wines". The same opinion was held by Peter the Great of Russia, who had a permanent purchasing agent in the Tokaj region and assigned a troop of fully armed grenadier guards every year to escort to St. Petersburg oxcarts loaded with *göncs* of Tokay.

Pope Benedict XIV, who received from Queen Maria Theresa of Hungary and Bohemia a gift of Tokay wine, went as far as to thank her with a versified pun:

Benedicta sit terra quae te germinavit
Benedicta mulier quae te misit
Benedictus ego qui te bibo

(Blessed the land that has grown thee, blessed the woman who has sent thee, and blessed am I [Benedict] who drink thee.)

AROUND THE "HUNGARIAN SEA"

Hungary is a small country. It produces three times less wine than its big Soviet brother, but in quality Hungarian growths are far superior.

A WINE BUYER entering a Tokay cooperative. *(Photo Chapman)*

Without leaving the shores of the Hungarian sea—as the Magyars call their Lake Balaton—a number of Hungary's most famed wines can be tasted on their home grounds, beginning with the Badacsonyi hills, covered with terraces of vineyards. Here, from the terrace of a tavern that was formerly the home of the poet Sandor Kisfaludy, one can sample wines within sight of their place of origin. The most notable are the white Badacsonyi Kéknyela (Blue Stalk) and Szürke Barát (Gray Friar), made from plants of Pinot Gris originally brought from France but yielding here an entirely different, aromatic dessert wine. Further northeast along the shores of Lake Balaton is an old fisherman's inn, Baricska Csarda, in the Balatonfüred-Csopak wine-growing district, where the traveler can taste the local version of the *Halaszle*. This is a traditional Hungarian fish soup made with pike, bleak, shad, or perch, and seasoned with red onions and paprika. *Halaszle* can be profitably accompanied by draughts of cool Olasz Rizling, from Italian Riesling grapes, or the light white Furmint of Balaton.

To the south the village of Mór cultivates a grape, the Ezerjó, whose name means "one thousand boons." A white wine, the Mori Ezerjo can be very sweet in good years, when the grape contains too much sugar for all of it to turn into alcohol. The fair white and red wines of Sopron are reputed for their medicinal value; they are, according to various claims, effective against plague, jaundice, fever, scurvy, apoplexy, epilepsy, swooning, dropsy, sciatica, and many other ailments. Szekszárd, in Transdanubia—the hilly region between the Danube and the Austrian border—is the home of Hungary's red wines, comprising 40 percent of the total production. Nearly one fourth of all Hungarian wines are reds made with the Kadarka grape, originally from Albania.

"Bull's Blood"

Outstanding among the Kadarka wines are those of the beautiful town of Eger, set in a broad, fertile valley in the mountainous country some 120 miles northeast of Budapest. Here there took place one of the first settlements of the Magyar tribes when they entered Hungary over a thousand years ago. Wine has been made in the Eger district for centuries, and the best, and best-known, is Egri Bikavér, or "Bull's Blood from Eger." Made chiefly of Kadarka mixed with Gamay, Cabernet Franc and other plants of French origin, Bull's Blood is dark red, almost pitch-black, full-bodied and aromatic, with a harshness mellowed by time. Much Egri Bikavér is consumed on the spot, either by the natives or by tourists wandering through Eger's hewn-in-the-rock cellars; but production is abundant enough for the government to export it throughout the world. Egri Bikavér is particularly suited to accompany some of the spicier Hungarian dishes, whose taste would overcome lesser wines—for example, the *fatanyeros,* a mixed grill of beef sirloin rubbed with pepper and red onion and grilled on embers; the *paprikas,* which can be made with just about anything (chicken, veal, pork, mutton, hare, deer, mushrooms, cauliflower), provided the basis consists of red onions and red paprika, and is doused with sour cream; or *kolozvari rakkotkaposzta,* the everyday dish of layered and stuffed cabbage.

VINEYARDS along the Elbe Valley, Czechoslovakia. *(Photo Ginette Laborde)*

The city of Eger is dotted with mansions and palaces of varied styles, mostly baroque, and below its streets there are several miles of underground passages and storage rooms. A viticultural station has been set up near the town and given the task of developing the vineyards and selecting new plants that might be suitable for the region. It might be feared that the wines of Hungary would become somewhat depersonalized from too harsh mass production, but this is unlikely, for the Hungarians are independent-minded and would not stand, one trusts, for that kind of nonsense.

BULGARIA

Bulgaria produces some 50 million gallons of wine a year, most of it of very ordinary quality, although there are some happy exceptions. Hardly any is known in the West, since most of the exports go to the Soviet Union, Czechoslovakia, East Germany, and other friendly neighbors. Viticulture is socialized, each village forming a collective farm, but the farmer can keep one fifth of his harvest. All production and exports are controlled by the state-run *Vinprmo* in Sofia.

A mountainous country, popular with skiers who do not fear bitter cold, Bulgaria has hot but brief summers. Hence the grape harvest cannot, as a rule, wait beyond October. If there is a shortage of laborers, the Bulgarian army is assigned special duty to help the growers, and children are let out of school to give a hand.

One of the most widely used grapes is the Gamza, which yields dark, full-bodied wines, sometimes proudly (and inaccurately) compared to French Burgundies. The best Gamza wines come from the Kramolin district at the foot of the Rhodope mountain range bordering Greece. The Mavroud grape gives an even deeper, rougher and more acid wine, which should be aged at least three or four years. It gains by being served with game, such as hare, deer, boar, and wild pig. The Pamid grape yields an ordinary light table wine, particularly appreciated in Sofia when made into rosé, while the Melnik variety is made into red wines that are usually too astringent for Western palates.

Three types of white wines stand out. One is made of Czervan-Misket, a pinkish muscat grape which gives a white wine with a slightly greenish tinge, particularly good when it comes from the district of Karlovo. A plainer, more ordinary white wine is made from the Dimiat, growing chiefly along the Black Sea coast of the "Bulgarian Riviera." A third plant, the Euxinograde, produces the country's best white wine.

Sparkling wines called "Bulgarian Champagne" are made according to the traditional French method, but a "continuous flow" Champagne factory, Soviet style, is being completed.

RUMANIA

Rumania, which makes twice as much wine as Bulgaria, also exports almost exclusively to the East. Much of the wine, however, is drunk on the spot. Hardly a bottle goes west, but Westerners should not be overly concerned, for few of the wines are worthy of interest. During World War II Rumania lost much of her wine-growing territory: Bessarabia, on the Moldavian plain, and Bucovina, were ceded to the Soviet Union, while the land south of the Danube went to Bulgaria.

Most Rumanian wines are white, and many are sweet. Among the best white wines are the Nicoresti, Mucel, Diosig, Murfatlar, Dealul Mare and Obobesti-Panciu Vârtiscoi. One red wine, Sarica-Niculitel, also stands out. The Cotnari winery, near the town of Iashi next to the Soviet border, is one of the country's largest, and also produces some of the best wines.

A genuine effort is being made to upgrade the quality of wines: many fine grape varieties are being tried out, and hybrids torn out. Small quantities of good Sauvignon, Cabernet and Furmint wines are being produced.

In summer, Rumanians like to drink their white wine mixed with soda water, and an interesting local custom is to make a refreshing beverage by macerating absinthe leaves in new wine, either white or slightly rosé. The resulting beverage is drunk ice-cold, either as an aperitif or during meals.

CZECHOSLOVAKIA

Czechoslovakia, created in 1918 by the splitting up of the Austro-Hungarian Empire, is more a land of beer than of wine. A few pleasant, light wines, chiefly white, are produced in Bohemia, Moravia, Slovakia, and the Carpathians. Prague has a number of centuries-old cellar taverns, mostly for beer, in which the state-appointed manager will be pleasantly surprised to be asked for wine, and might produce an interesting bottle.

Since World War II, the quality of wines has been improving; collectivization of tiny vineyards has helped set up quality standards that were practically nonexistent before. The best wines—red or white, still or sparkling—are made near Mělník, in the Elbe valley north of Prague.

YUGOSLAVIA

Yugoslavia was created in 1918 by uniting wine-growing Serbia and Macedonia with such choice viticultural southern Slavic countries as Croatia, Slovenia, and Voivodina, previously part of the Austro-Hungarian Empire. A complex mosaic of peoples, cultures, and traditions, it is also varied geographically. Wild here, well ordered there, drearily monotonous in parts, magnificently beautiful in others, it has features, climate—and wines—for virtually every taste.

Yugoslavia has a rich wine-growing tradition, dating back to the time of ancient Greece. After World War II, wine growing was encouraged by the government of Josip Broz Tito. Realizing that a rigidly socialized mass production would not serve the nation's economy as well as concentration on growing quality wines, Tito's government built up a system that mixes state-owned viticultural combines with agricultural cooperatives and small private vineyards. The result has been to raise Yugoslav wines to a creditably high level. Today more than half the wines exported are bottled, rather than being shipped in bulk to disappear into the anonymity of blending vats. Selection of vines is closely supervised by several specialized institutes, and nearly all of the finer or "noble" European vines are either in production or being tested.

The country can be roughly divided into three wine-growing regions: The North, the East, and the Adriatic coast. These regions surround a central, wooded area, chiefly Bosnia and western Serbia, which produces no wine but makes up for this lack by production of Slivovitz, Yugoslavia's fiery plum brandy.

TENDRIL AND GRAPE CLUSTER MOTIF, from a bas-relief on a Bogonid tomb in Bosnia, Yugoslavia. *(Photo Georges Viollon-Rapho)*

Opposite: WICKER-BOUND WINE BOTTLES containing the wine of Korcula, Yugoslavia. *(Photo Goldner)*

THE NORTH

Yugoslavia's northern wine-growing region consists of some 200,000 acres of vineyards stretching along the Austrian and Hungarian borders. It produces mostly white wines. A wide variety of vines, both domestic and imported, is used, including Franken Riesling, Pinot Blanc and Gris, Sauvignon, Furmint, Veltliner, Chasselas, Johannisberg Riesling, Gewürztraminer, and the local Plavetz, among others. Slankamenda and Dinka, also local plants, are extensively used to make both white and rosé wines.

Some of the region's best wines come from mountainous slopes near the Drava and Mura rivers, around the Slovenian town of Maribor. This is particularly true of the small amounts of dry fruity white wine bottled "at the estate" by the Maribor Wine Cooperative. Another excellent white wine is the Traminer produced in Radgona, the northernmost Yugoslav vineyard, in a mountainous niche bordered by Austria and Hungary. Some fine Sylvaners and Traminers are also produced in the district of Jeruzalem, near Ormoz.

Nearby, but across the provincial boundary in Croatia, similar wines, although perhaps with less distinction, are made around the town of Varazdin. They are light and pleasant everyday wines. Farther east in Croatia, around the towns of Slavonski Brod, Erdut and Ilok, better and stronger wines stem from Riesling, Pinot Blanc, Sémillon or Sauvignon grapes. Rosé wines are made only in limited amounts, the best-known being the slightly acid and light Cvicek (pronounced Tzvichek) made chiefly from the Kölner Noir grape, grown in Slovenia along the Sava River and in Croatia along the Danube. There are a few red wines, the best coming from Pinot Noir, Merlot, and Kadarka grapes. Good red wines are produced near the Hungarian and Rumanian borders, in the regions of Subotica and Vrsac. Further south, near the Danube east of Belgrade, the hills of Fruska Gora are reputed for their fruity, sometimes slightly sweet wines.

THE EAST

Yugoslav's eastern wine-growing region, with some 300,000 acres of vineyards, stretches through Serbia from about the level of Belgrade, down through Macedonia, to the Greek border. Red wine predominates, much of it from the Prokupac grape, yielding mostly ordinary table wine of little interest, although some excellent wines as well. The better wines generally come from the new state-owned vineyards planted with Pinot, Gamay, Merlot and Cabernet varieties.

The best-known come from the huge Vancacka wine cooperative in northern Serbia and from the Aleksandrovac and Zupa districts in central Serbia. Of special interest is the Zupsko Crno, a strong, full-bodied, tannin-rich wine (its name comes from *cerno,* meaning black, except in the case of wine, when it means red). In southern Serbia, the red Prokupac and Plovdina wines are heavier, but not so heavy as the Macedonian wines, which tend toward Oriental sweetness. Many come from the Kavadarci winery in central Macedonia, one of the most modern in Yugoslavia. Rosé wines *(Ruzica),* produced in the Zupa and Vlasotince districts, are made chiefly from the Prokupac grape.

White wines are also made in this eastern region, notably with the Smeredevka grape, along the Danube east of Belgrade. They are fruity, usually quite strong, and are often drunk in summer mixed with sparkling water.

THE ADRIATIC

By all odds, the most charming wine region of Yugoslavia is the Adriatic, where some 160,000 acres of vineyards run from Istria, just south of Trieste, along a thin strip all the way to the Albanian border, and include some 1,000 offshore islands. This is the fabled Dalmatian Coast, site of one of the most spectacular tourists booms since the 1950's.

Many archaeological findings testify that this region has been a wine-growing area for more than 2,000 years. Greek-style wine vessels unearthed on the island of Vis; a statue in Split of a girl bearing a *kantharos* (large goblet) of wine; bacchantes carved in stone in the ruins of ancient Salonae, once capital of Dalmatia, are but a few examples.

Like its Italian neighbor across the Adriatic, Dalmatia and other parts of this region have an abundance and variety of plain, fair, and good wines, "southern style," and also a few excellent ones, apt to satisfy the most demanding visitor, if not the wine snob. Western European grape varieties are used, as are many indigenous plants that cannot be found elsewhere, such as Kuyundjusha, Bogdanusha, Slataritza, Medna, Marashtina, Vugava, Gerk, Kerkoshia and Bena for white wines, and Plavina, Babic, Platatz Mali, Okatatz, Vranatz, and Hervatitza for red and rosé wines.

Red wines are predominant in this region: One of the best known is Krashki Teran, produced from the Refosco grape on the Slovenian littoral near Trieste—and much in demand in Italy. It is fresh and has a high acidity, making it an excellent summer wine. Merlot and Cabernet wines are made in the Brda area, where one of the best-known wines is Château Dobrovo, from the Dobrovo win-

Left: BACCHANTE.
Roman bas-relief
found at Solin near Split,
Yugoslavia. *(Photo Marcel Jelaska)*

Right: TERRACED VINEYARDS on Vis Island, in the Adriatic Sea.
Of unknown origin, and pre-dating the ancient Greeks,
they are still cultivated today. *(Photo Marcel Jelaska)*

ery. Istria produces good red wines from the Refosco (or Teran), Gamay (or Borgonja), and Barbera varieties.

The islands of Krk (pronounced *kurk*), Cres, Losinj and Susak, and northern Dalmatia facing them, produce mostly common wines. Toward Sibenik and further southeast in central and southern Dalmatia the wines assume an increasingly southern character, becoming fuller, deeper in color and richer in body. Some of them come from the best Yugoslav varieties of grape, and from the best sites, such as the Peljesac peninsula and the southern slopes of the islands of Brac, Hvar, Vis, Bisevo, Korcula, and Lastovo. The best quality red wine from the Plavatz variety is produced on Peljesac and is known as Dingac (pronounced *Deengatch*). It is the first Yugoslav wine to have been subjected to strict quality control, and most of it is exported. Dingac has a dark red color with violet reflections, and a typical, full-bodied flavor along with a slight sweetness and astringency—all these characteristics being balanced in a fresh and harmonious whole. In central Dalmatia, great amounts of strong, deeply colored wines are produced chiefly for export to Western and Eastern Europe for blending.

The principal white wines in this region are produced in Istria, along the Slovenian littoral, in northern Dalmatia and on several islands, as well as in Herzegovina. Starting northwest on the Slovenian littoral, the Vipava valley district produces strong, smooth white wines with a pleasant green tinge, known as the Vipavchan wines. In Istria, there is the well-known Malvasia Istarska, and Sémillon and Pinot wines.

Northern Dalmatia produces good table wines from Debit and Trebbiano varieties. In other parts of Dalmatia, three well-balanced, pleasant, greenish-yellow table wines, are produced: Kujundjusha from Imotski, Hvarski Bijelo from the island of Hvar, and Visko Bijelo from Vis. The Vugava, also from Vis, is one of the best white wines.

Numerous rosés are produced in Dalmatia, especially around Split and on Vis; many of them are rather strong, sometimes almost red in color.

Among sparkling wines in Yugoslavia, the best is that of Gornja Radgona, in northeastern Slovenia; it has been produced there for 150 years following the classical Champagne method. A special type of sparkling wine, Bakarska Vodica ("blessed water of Bakar"), is produced in Kvarner on the Croatian littoral.

Yugoslavia produces a large assortment of dessert wines, of which the "prosek wine," made in Dalmatia from the Marashtina grape, deserves a mention.

Food in Yugoslavia is as varied as the wines, and there are dishes for every taste: Hungarian-style *paprikas* and *gulas* (goulashes) in the north; Austrian-type cooking in the west; Italian-type cuisine with olive oil, fish, vegetables, and cheese (notably the *Paski* from Pag and *Livanjski* from Livno) along the Dalmatian coast; and outright Oriental, Greek, or Turkish, in the east and southeast—lamb or beef on the spit, and pilafs, usually spicy, rather fatty, and chockful of calories. Several specialties can be found just about anywhere. They include goulashes and *Wienerschnitzel* among foreign ones, and two national dishes: *cevapcici*, bits of chopped meat grilled with onion, pimento and herbs—a sort of Yugoslav hamburger; and *raznjici,* chunks of meat with assorted condiments cooked on spits similar to Georgian shashliks—a counterpart of sorts to the hot dog.

A FEW YUGOSLAVIAN WINES. *(Photo Marcel Jelaska)*

United States
Latin America
South America
Australia

A PRE-PROHIBITION American wine label,
used for wine which had been baptised "Sauterne." *(Photo A. Sponagel)*

Preceding page: WINEGROWER tying vine to a support
in the Paul Masson vineyard in California. *(Photo Paul Masson)*

UNITED STATES

The first association of America with wine was made in the *Greenland Saga,* a Nordic family chronicle of the 12th century. This chronicle tells about the exploits of Leif Ericson, a Norse mariner and son of Eric the Red, discoverer and uncrowned king of Greenland.

Just before the turn of the millennium, Leif and a crew of seafaring adventurers landed along the New England or Canadian coast, near the mouth of a river. That evening, one of Leif's crew, Tyrker, the only German of the expedition, disappeared. Hours later he returned in great excitement and explained: "I did not go much further than the others. But I made a great discovery: I found vines and bunches of grapes."

Leif Ericson gave this land the name of Vinland ("wine land") by which the American continent was known in Norse literature for centuries.

As European *Vitis vinifera* cannot easily survive the rigorous climate of the American Northeast, and as no wild *vinifera* was ever discovered there, there is no doubt that the plant found by Tyrker was a native species, probably *Vitis labrusca,* the same that explorers centuries later saw crowding the forests of the North Atlantic coast. Some of the early colonists made wine with *labrusca* grapes, but it was not very palatable and they turned to importing the proven winemaking varieties from Europe.

The Spanish conquerors in the west were followed by Jesuit and Franciscan missionaries, whose task was to convert the natives—and they needed sacramental wine. The Jesuits settled chiefly in Mexico, but the Franciscans, led by Father Junipero Serra, went on northward. In 1769, Father Junipero established the Mission San Diego, where he planted European vines and discovered that California was a winegrower's paradise. The first California vine became known as the "Mission Vine"; it was certainly a European *vinifera,* perhaps an accidental offshoot whose ancestors have remained anonymous.

The first professional winegrower in California was a Frenchman, precursor of many who brought their traditional skill to the New World. Born in Bordeaux with the appropriate name of Jean Louis Vignes, he settled at the pueblo of Los Angeles and planted vines on 104 acres in the heart of what is now downtown Los Angeles, just west of the Los Angeles River. Vignes planted much Mission grape, but he was also the first to import and experiment with new European varieties, and the first to age his commercial vintages.

THE WINE-GROWING COLONEL

One of the greatest contributors to California wine making was an exiled Hungarian nobleman, Count Agoston Haraszthy, a former colonel in the Royal Guard under Austrian Emperor Ferdinand. Haraszthy sailed to the New World in 1840. First settling in Wisconsin, he trekked across the country during the Gold Rush, and engaged in numerous more or less successful ventures in agriculture, politics, cattle raising, and gold assaying and refining.

When the colonel (he preferred this more democratic title than that of count),

AGOSTON HARASZTHY'S RESIDENCE, the Buena Vista Ranch, Sonoma County, California. *(Photo Wine Institute)*

accompanied by his wife and two of his sons, reached California in 1849, they first settled in San Diego, then a blossoming city of 650 inhabitants. With his thick black hair, black mustache, black beard, and cane in hand, Haraszthy became a well-known, imposing figure in the pioneering community. In 1852, he purchased more than 200 acres of land near San Francisco's Mission Dolores, named his new domain Los Flores, and immediately started building a stately house and planting vine cuttings that he had requested from his native Hungary —among them, presumably, the Zinfandel, one of the most widely used vine plants in California today. Haraszthy later moved to the Sonoma Valley, where his land, which he called Buena Vista, gently sloped to the foothills of the Mayacamas Hills to the east. The property was not far from the vineyards of General Mariano Vallejo, the last Mexican military commandant of Sonoma and another pioneer of American wine growing. Until the colonel's two sons married the general's two daughters, a brisk rivalry existed between the wine-growing neighbors, and their efforts to surpass each other certainly gave a healthy boost to California's wine industry.

By the end of 1858, Haraszthy claimed he had 165 varieties of foreign vines growing at Buena Vista. By 1861, he was California's authority on wine and president of the State Agricultural Society. That year he traveled to Europe, where he purchased some 100,000 European vine plants. Then, at the height of his wine-growing career, his luck soured. He became involved in both political and commercial difficulties. Having spent less than twenty years in California, and having covered a vast area there with vines, Haraszthy decided to seek his fortune elsewhere. He acquired a large plantation in Nicaragua and started the cultivation of sugar cane and construction of a distillery. On July 6, 1869, Agoston Haraszthy disappeared, never to be seen again. There was an alligator-

THE LACHRYMA MONTIS VINEYARDS, home of Gen. Mariano Vallejo, Sonoma County, California. *Engraving. (Photo Wine Institute)*

infested stream on his property. Some believe he was killed or drowned while attempting to cross it.

WINE GROWING—BEFORE AND AFTER PROHIBITION

The vineyards the colonel had done so much to help propagate flourished. Sometime in the late 19th century the phylloxera reached California. The disease was troublesome, but thanks to the fertility of the soil and the technique of grafting it was not as destructive as in Europe. Interest in wine growing was strong and the industry continued to grow.

On the eve of World War I, the American people were drinking more than 50 million gallons of wine a year, most of it produced domestically. Then yet another disaster threatened the flourishing industry. In 1917, the Eighteenth Amendment to the Constitution, prohibiting the manufacture and sale of intoxicating liquors, was passed. It was ratified by thirty-six state legislatures by January 16, 1919, to become effective a year later. This was followed by the Volstead Act, adopted by Congress in spite of President Wilson's veto in October 1919, which defined intoxicating beverages as those containing one half of 1 percent or more of alcohol.

The resulting chaos, with its bizarre intermingling of everything from the comic to the criminal, is a well-known part of American folklore. Suffice to say that it took public opinion only a few years to turn against prohibition, which was repealed in 1933.

The repeal of prohibition was not immediately followed by an improvement in the quality of American wines. In fact, the contrary occurred. Repeal created a large demand which the newly restored industry could not fulfill. The first post-repeal vintages were of extremely poor quality—not so much because of conscious fraud as because a whole generation of wine makers had been lost and the special techniques for overcoming the problems of wine making in a hot climate had to be relearned. California's winegrowers were the first to take stock of this difficult situation and to promote official quality standards. Federal regulations establishing nationwide standards followed in 1936, marking the beginning of the renaissance of American wine making. Since 1937, the quantity of wine produced in the United States has continued to increase, and the quality has improved. Today, the United States holds eighth place among the world's wine producers, although it is really not a consumer. An American drinks on the average about a gallon of wine a year, as against more than thirty gallons for the average Frenchman.

CALIFORNIA WINES TODAY

California produces 85 to 90 percent of all American wines. At first California growers gave most of their wines European names, such as Burgundy or Chianti, since to give them the names of Californian villages or townships of origin would have been confusing. Then growers slowly turned to using the name of the principal grape variety used to make a wine, thereby adopting the "varietal" form of nomenclature. It was a courageous move, since it meant introducing to consumers names with which they were not at all familiar, such as Pinot Noir, Cabernet, and others. For these wines, the label also carried the name of the vineyard or region or both, with the result that some of the good American wines finally took on an identity of their own. A good example is "Napa Barbera, L. M. Martini," indicating that the wine is produced in Napa County, chiefly from the Barbera plant (which originated in Italy), by Louis Martini.

Of course many wines still carry the semigeneric names, such as Sherry or Burgundy. These may never be completely eliminated, though the wines are far removed from the European growth whose names they borrow, and offer no valid yardstick for comparison.

The northern California coastal regions centered around San Francisco produce, as a rule, the best table wines. From north to south, the most important wine-growing regions here are the counties of Mendocino, Sonoma, Napa, Alameda and Santa Clara, the latter extending into Santa Cruz and San Benito counties.

A visit through the California wineries, many of which are hospitably open to the public, can be an edifying experience. Almadén, some 60 miles southeast of San Francisco, is, for example, typical of the American tradition of wine making. Almadén is the name of a former mercury-mining village (itself named after the mercury mines of Almadén in Spain). The vineyard was founded in 1852 by two Frenchmen, Etienne Thée, a farmer from Bordeaux who was lured to California at the time of the Gold Rush, and Charles Lefranc, a Parisian tailor who followed Thée and married his daughter. It was developed and greatly expanded as a post-repeal winery by San Francisco businessman Louis Benoist, but has recently been absorbed as a subsidiary by one of the big whiskey concerns. As the original vineyard on the western rim of the Santa Clara Valley has been invaded by housing developments, Almadén now includes an area around Paicines, in San Benito County, to the south, as well as vineyards in the Sonoma and Livermore valleys to the north.

Though on a latitude about 600 miles south of Bordeaux, Paicines vineyards comprise one of the coolest viticultural spots in California. Situated 600 to 1,200 feet above sea level, 20 miles east of Monterey Bay, they are classified among the Number 1 and Number 2 regions. These are the coolest ones, hence the most favorable for growing fine wine-bearing grapes; the hottest, and least favorable, are classified as Number 5.

The finest plants in Region 1 vineyards mature slowly, and are never gathered before October, and sometimes as late as November or December. Paicines soil seems particularly well suited to the Pinot Noir, one of the most distinguished of all the French plants. Almadén's vineyards boast hundreds of acres of Pinot Noir—more than anywhere else in California. The grapes are harvested into metal containers and loaded into wagons which are then pulled by tractors across the hills to the wineries at Cienega for red wines and Paicines for whites.

Passing between two old wooden, screw-type presses standing on both sides of the entrance in Cienega (and still occasionally used) the grapes for red wines

TREADING AND PRESSING GRAPES in California.
Notice the use of Chinese labor, originally imported for work on the transcontinental railroad. Wood engraving by P. Frenzeny, 1850. *(Photo Wine Institute)*

enter the domain of the cellar master. The juice of the grapes is extracted with the modern Willmes press, a perforated, horizontal steel cylinder against which grapes are squeezed by a pneumatically expanding rubber tube in the center. The "must" is pumped into immaculate, glass-lined concrete vats lined up on both sides of an overlooking gallery. Here it ferments before being racked into large redwood casks, where the wine waits for the lees to be deposited. After clearing, it is racked once more into puncheons, the traditional, 160-gallon California wine barrels. The winery and storage rooms across the road are air-conditioned, a necessity in California, for Indian summer can turn October into the year's hottest month.

The Almadén vineyards produce a complete selection of so-called "premium wines." White wines, like the reds, are usually sold under the varietal label: dry Sémillon, excellent Pinot Chardonnay and Pinot Blanc, the Johannisberg Riesling (a true Riesling), the fruity and spicy Traminer, the Sylvaner, and the Gray Riesling (not really a Riesling). There is also the popular Grenache rosé.

The vineyards offer an assortment of "Champagne" sparkling wines, bottle-fermented. Among them is the Almadén Brut, which is, in fact, a clean dry *mousseux*, issuing in large part from the French Pinot Chardonnay grape.

That part of Sonoma Valley which the Indians called the Valley of the Moon is

the home of the Buena Vista wineries, formerly the domain of Colonel Haraszthy. The original winery and many of its wine cellars were destroyed by the great earthquake of 1906, and it is said that some of the Haraszthy champagne may still be buried under the rock. Buena Vista produces much Zinfandel, red wine from the plant Haraszthy was so proud of, with a bright ruby color, a special fruitiness and a flavor that has been compared to that of raspberries. There is an estate-bottled Cabernet Sauvignon varietal red, and several estate-bottled whites: Pinot Chardonnay, Riesling, Johannisberg, Traminer, Sylvaner, and Sonoma Sémillon, and a palatable Pinot Chardonnay "Champagne" Brut.

Is It Italian? Is It Swiss?

Further north, in the town of Asti, there stands one of the best known, biggest, and most visited Californian wineries, Italian Swiss Colony.

Its name comes from the Italian Swiss Agricultural Colony founded in 1881 by Andrea Sbarboro, a prosperous Italian-American grocer, for the purpose of helping Italian and Swiss emigrants settle in the West. After many tribulations, Sbarboro gave up the colony idea and started producing and selling wine. The shifting ownership of what is now a huge enterprise specializing in medium- and low-priced wines would be a story in itself.

Opposite page:
CALIFORNIA
VINEYARDS.
(Photo Paul Masson)

TRANSPORTING
GRAPES
in California.
Lift trucks are used
to carry the clusters to
the vintners' wagon train.
(Photo George Knight)

Some of the labels on Italian Swiss Colony wines have been found most perplexing by foreign visitors, who wonder why a single label gives the impression that the wine comes from as many as four countries at once: Italian Swiss Sparkling Burgundy, is, after all, an American wine. So is the Italian Swiss Paree, a lightly sparkling wine whose label, however, cannot legally hint at the presence of a single bubble of carbon dioxide, since this might identify it as "Champagne-type." Such wines with minimal CO_2 pressure are sometimes referred to as "table wines plus" (meaning plus CO_2). The Burgundy and Paree can probably best be left on the shelf, but the winery's Vermouths (Lejon and G.&D.) are sound merchandise, as is the rosé wine.

The Napa Valley wine-growing region stretches between San Francisco Bay north to the 1,800-foot-tall Mount St. Helena. The region, which includes Napa County and some of Solano County, is America's best-known wine-growing district, and has a striking beauty of its own. Its vineyards stretch toward wooded hills and snow-capped mountains from which cool streams run down to the Napa River and San Pablo Bay. The district produces some of the best California wines—varietals such as Cabernet Sauvignon, Pinot Noir, Pinot Chardonnay or Chenin Blanc, and quantities of good, solid wine, mostly red, that winds up under such semi-generic names as California Burgundy and California Chianti.

Some of Napa's finest wines are produced by Louis Martini of St. Helena, a traditional wine maker who scoffs at mass production, carbonation, flavoring, and other such sophistications. A grizzled native of the Italian Riviera, Martini produces top varietals, notably from a vineyard named Monte Russo (from its volcanic red soil) situated at an altitude of over 1,000 feet at the crest of the Mayacamas Hills which divide Napa and Sonoma counties. Martini's Mountain Wines range from the Mountain Red Burgundy to more distinguished varietals such as Mountain Pinot Noir, and the scarce, pale yellow Mountain Gewürztraminer. Fruity, fresh and spicy, the latter has a pleasant, light sweetness, and is great with seafood from the Bay.

Among other leading winegrowers in the valley, Beringer Brothers in St. Helena produce worthy Beringer Private Stock wines and a Pinot Noir blend which goes under the name of Berenblut ("bear's blood"). The Charles Krug Winery is noted for its fruity Gamay Beaujolais and for several of its dry whites. Inglenook Vineyard Company (only lately absorbed by one of the big combines) is probably best known for its fine Cabernet Sauvignon and Pinot Noir wines. Beaulieu Vineyard (B.V.) in Rutherford, founded in 1900 by Frenchman Georges Latour, is now headed by André Tchelistcheff, dean of California's wine makers. It has a good range of current California varietals with Cabernets and Pinot Noirs sold under a Private Reserve label after having been aged in the bottle for several years. It also has very limited production of an excellent "Champagne" made exclusively of Pinot Chardonnay. High on the eastern slope of the Napa Valley are Leyland Stewart's Souverain Cellars. His fine red and white wines are well known in California but rarely seen elsewhere.

THE "CHAMPAGNE" CELLARS of Paul Masson in Saratoga, California.
Note the traditional California barrels or "puncheons."
Made of redwood, they hold 160 gallons. *(Photo Ansel Adams and Pirkle Jones)*

The Christian Brothers

In the lower Napa Valley are the famed Mont La Salle vineyards of the Brothers of the Christian Schools, more commonly known as the Christian Brothers, a Roman Catholic society founded in France around 1680 for the purpose of educating young men and boys. The Brothers came to America in the 1800's, and today run more than a hundred institutes in the country—notably Manhattan College in New York City and La Salle Military Academy on Long Island. They started making sacramental wine in California in 1879. The demand grew, and today the society runs a brisk business.

Wines are made in stone wineries built in 1903 at Mont La Salle, the Brothers' novitiate at the edge of the valley. Wines to be aged are then taken down to the St. Helena cellars, one of the largest networks of stone cellars in the world, and then returned to be bottled at Mont La Salle. The Brothers produce more than half a million gallons of table and sacramental wines a year, as well as thousands of cases of undistinguished sparkling wines, made by the bulk process in the St. Helena cellars. Table wines, aperitifs and dessert wines are sold under the Christian Brothers label, while sacramental wines are called Mont La Salle. Profits go to the novitiate and to run several schools near San Francisco.

In Alameda County, around the city of Livermore some forty miles southeast of San Francisco, lies the Livermore Valley. Its gravelly soil and climate provide fine conditions for the European vine. Here the emphasis is put on white table wines—from Sauvignon Blanc, Sémillon, Pinot Blanc, Gray Riesling and Chardonnay plants. The Cresta Blanca vineyards in Livermore, founded in 1883, are named after a nearby white limestone crest. A few years ago, with assistance from oenologists at the University of California, they began producing a sweet wine made from the Sémillon grape of the French Sauternes region, artificially infected with *Botrytis cinerea* mold. This is the same tiny mold that withers the grapes producing Château Yquem and other sweet wines of Sauternes. The grapes with this mold are selected individually, yielding an unusual wine, sweet and fruity, the Premier Sémillon. The Cresta Blanca Vineyards also produce white table wines, as well as reds from grapes grown in the Napa Valley. Sparkling wines are bottle-fermented, including the usual range of "Champagnes" and the almost inevitable "Sparkling Burgundy." The Flor Sherry can be quite good.

Wente Brothers, also in Livermore, was founded by Carl Wente, a German-Swiss immigrant, in 1883. He specialized in white wines, producing some of the best in California. Still a family business, Wente Brothers excels in the Pinot Chardonnay and Sauvignon Blanc varietals. Some of the plants in the Wente vineyards came directly from Château Yquem in the Bordeaux area.

Concannon Vineyard in Livermore, founded by James Concannon, an Irish immigrant, in the 1880's, is now run by his two grandsons, Joseph and James. Like its neighbor Wente, it emphasizes whites and produces an assortment of varietals made chiefly from Sauvignon Blanc, Sémillon and other superior plants.

Santa Clara County, once a flourishing vineyard, is becoming a suburb of San Francisco and the grape is retreating south. The Paul Masson Winery is known

"TURNING" CHAMPAGNE. Wire masks are worn to prevent injury in case of an explosion. *(Photo Wine Institute)*

for its "Champagnes," and particularly for a 67-foot tall "Champagne fountain" over a reflecting pool and a spiral ramp reaching to the visitors' gallery. The "Champagne Brut" is available in standard bottles, double-sized magnums, and quadruple-size Jeroboams. Paul Masson has the usual full range of wines, including a red which has the distinction of having been rated by French experts as the best American wine of a dozen put on sale in a few French supermarkets in 1966. Nearby lie the vineyards of Martin Ray, a stockbroker turned winegrower. Ray once owned the Paul Masson Champagne Company, but now makes his own excellent bottle-fermented sparkling wines.

The largest wine producing area in California is the San Joaquin Valley, running southeast from Sacramento through San Joaquin, Stanislaus, Madera, Fresno, Tulare, Kings, and Kern counties. In the fertile, irrigated valleys of this region, much of the yield goes to make fortified wines, which as a rule are lacking in interest. Exceptionally, the Ficklin Vineyards on the outskirts of Madera specialize in growing true port varieties of the Douro region from Tinta Cao and other grape varieties. Their best wines are comparable to Portuguese Douro ports. In recent years, northern San Joaquin Valley has become the largest table wine producing region—over 25 million gallons annually. This is the main source of California's "standard" wines—well made, drinkable, unexciting, and bearing such names as Viva Vino di Pietri, Gallo's Paisano, Roma's Barberone, and of course Zinfandel.

The third and last wine-growing region in the west is actually a string of several districts in Southern California, starting east of Los Angeles with San Bernardino County, and extending to around San Diego, where Father Junipero Serra planted the first Mission wines two hundred years ago, but where little wine is made today. The best Southern California wines come from the Cucamonga district, in southern San Bernardino County.

EASTERN AMERICAN WINES

Across the continent, the State of New York, which produces less than one tenth as much wine as California, is the other American wine-growing district of importance. Its vineyards are unique, for until recently the grapes almost exclusively used for wine have been of the *Vitis labrusca* species and its more or less legitimate offspring (*Labrusca* wines are also grown in Washington, Ohio, Michigan, Missouri, Canada and Japan).

The story of the *labrusca* grape and its offspring is a long one. For nearly two centuries, beginning with the early American settlers, stubborn attempts were made to grow European *Vitis vinifera* in the Eastern regions of America, but without much success. Many varieties of American native vine existed, but they were looked upon with disdain by the settlers. Thomas Jefferson is credited with being the first to encourage winegrowers to use the native *labrusca*. His own experiments were not a success. The real start took place when one John Adlum set out a vineyard in Georgetown, Virginia, in the 1820's. The Catawba grape, probably an accidental hybrid of wild local varieties, was cultivated there, and

SNOW helps wine age in the barrel.
New York State.
(Photo Widmer's Wine Cellars)

PRUNING VINES in winter
in upstate New York.
(Photo Taylor Wine Company)

Adlum was so satisfied with spreading its use that he let himself be carried away with enthusiasm. "In bringing this grape into public notice," he said, "I have rendered my country a greater service than I could have done had I paid off the national debt." The Catawba grape was followed by the Isabella, first grown on Long Island, New York.

At about the same time there appeared, in the gardens of one Ephraim Bull of Concord, Massachusetts, the seedling of an unknown variety of *labrusca*. It was a blue-black, slipskin grape, so called because the skin slips off with ease, while the pips remain firmly embedded in the pulp. Bull's Concord grape, an accidental cross, turned out to be a good eating grape, readily adaptable to almost any soil, and capable of withstanding almost any pest and weather. Thousands of acres of Concord grapes were planted in a few years, and a large number of variants were developed, giving rise to the New York State wine industry. Ironically, the Concord grape is, in fact, hardly suited for wine making. It has to be sugared abundantly, and its "foxy" flavor is obtrusive. (Early settlers called the native *labrusca* the "fox grape"—hence its "foxy" taste.) Yet it is used to make much New York State "Port" and "Burgundy."

At first, Eastern vineyards were planted haphazardly, but it soon became apparent that the best wine came from the Finger Lakes region of New York State. The reasons for this phenomenon cannot precisely be described, but are usually attributed to a soil rich in minerals, the climate-regulating influence of

the lakes, the hot summer days, and the ample snow in winter, which protects the vine roots from excessive cold.

Best-selling wines of New York State are probably the least interesting ones: they are the so-called Ports, Sherries, Sauternes, and Burgundies. While there is now a new surge of effort in the Finger Lakes region (as well as in other parts of the East) to grow European varieties of grapes, most of the wines in Eastern and Midwestern vineyards still come from the old standard native sorts. Whether "foxy" or fruity, all of these Eastern and Midwestern wines are altogether different from European or Californian wines.

An important current development is a shift away from the old *labruscas* to a range of new hybrid grape varieties. These new grapes represent an attempt to apply genetics to viticulture, just as it has been applied with such success to many food grains and fruits. By crossing certain native (but non-"foxy") species with the classic European-type *vinifera* varieties, the object is to obtain new varieties combining *vinifera* quality with the ruggedness required for survival and good production under adverse conditions. Thus far, results have been variable but have included some significant successes. The best of the new hybrids obtained from the experiments are opening up the possibility of serious wine growing in many parts of the United States where it has not been tried before. However, most European winegrowers doubt that great wines can ever be achieved from hybrids.

In the case of Eastern red wines, most are named generically, rather than by grape variety. In New York State, Gold Seal Vineyards, the Taylor Wine Company, and Widmer's Wine Cellars have fair wines of the so called Claret or Burgundy type. Taylor also has an interesting rosé, made from a relatively new French-American hybrid.

When it comes to white wines, varietals predominate—with that "foxy" taste, take it or leave it. Widmer's has a particularly broad range of varietals made from the Diana, Dutchess, Elvira, Moore's Diamond, Delaware, Salem, Niagara, and Vergennes grapes. Vergennes is an American variety, probably *labrusca*, dating back to Revolutionary days, and is well worth sampling.

New York State also produces sparkling wines, from assorted blends, including the usual range of "Champagnes," "Pink Champagnes," and "Sparkling Burgundies." The best ones are made with the Champagne method. Gold Seal Vineyards produces two "Champagnes": one is a standard product considered by some wine lovers to be lacking in sublety and too sweet; the other is produced under the aegis of a French wine maker, Charles Fournier, former Gold Seal president. This "Champagne" bears Fournier's name on the bottle, and is as good as any in the East. Gold Seal also markets a very pleasant semisparkling wine called Fournier Natur. Light and dry, it is very slightly carbonated.

The Pleasant Valley Winery (a subsidiary of Taylor), founded in 1860 and the oldest winery in New York State, produces the well-known Great Western "Champagnes." It has also recently launched a *Vin Blanc Sec* made of three French-American hybrids.

Perhaps the apostle of French-American hybrids in the United States is Philip Wagner, who with his wife Jocelyn runs a vine nursery and vineyard in Riderwood, near Baltimore, Maryland. The Wagners make red, white, and rosé wines, sold under the name of Boordy Vineyard. Most of the wine is sold in the Baltimore-Washington area, and much of it is delightful.

Wines somewhat similar to those of New York State are produced in Canada, almost exclusively on the Niagara Peninsula southwest of Lake Ontario. But the production is small and less varied. As might be expected, such wines as "Canadian Port" or "Canadian Sherry" are among the products, but there are also table wines—reds lacking in body, but a few delicately perfumed whites, some of them quite dry, ranging in color from pale yellow to dark gold.

THE OUTLOOK

If North America today produces many fair and a few excellent wines, the future of her wine-growing industry seems to be even brighter. The public, for one, is slowly becoming a wine-drinking public, and the number of connoisseurs is growing. Many Americans buy the most prestigious foreign vintages, France being, as one might expect, the greatest exporter of foreign wines to the U.S.

HARVESTING GRAPES and making wine in Pleasant Valley.
Engraving by Theo. B. Davis, Harper's Weekly, 1872. *Granger collection.*

SAMPLING PARTY.

JULES MASSON.

THE WINE-PRESS.

THE CHAMPAGNE CELLAR.

THE VINTAGE.

PLEASANT VALLEY DURING THE GRAPE HARVEST.

FINISHING SPARKLING WINE.

PLEASANT VALLEY WINE-CELLAR.

and Canada. But the foreign wine import, even though it has been constantly increasing in recent years, has been decreasing in average quality. Mediocre wines cross the borders and, either because of their snob value or because they are produced with lower-cost labor, compete with American wines, though they may not be superior to them. Nevertheless, Americans drank 175 million gallons of American wines in the year 1965–66, up 23 percent in ten years, and domestic wines still account for more than 90 percent of total consumption.

It seems that the solution is for American growers to continue producing an increasing amount of fine table wines, to compete at least with some of the less distinguished foreign imports. That such wines can be produced is no longer to be doubted. That they can be produced in large enough quantities and economically remains to be seen.

So far, nowhere in the world have excellent wines been mass produced, and many wine experts tend to mistrust such "industrial" producers as Ernest Gallo, the head of the E.&J. Gallo Winery of Modesto, in the San Joaquin Valley, California. This winery produces and ships many millions of gallons annually with the aid of such equipment as huge open-air steel tanks, steel pipes, powerful pumps and wine trucks and railroad cars. Mr. Gallo maintains that "We have found that it's practical for us to produce quality in quantity." This is possible, he once told *Time* Magazine, "by applying the same sound business principles to the wine industry that are being applied to any other American industry . . . If we use grapes equivalent to the best, if we have wine know-how equivalent to the best, and if we have facilities equivalent to the best, it must follow that our *quality* is equivalent to the best."

Yet, sound as such arguments appear, they just don't seem to work; there may be something of a mystery in wine, which requires unhurried devotion.

John Daniel, until recently owner of Inglenook Vineyard in California, has what may be a sounder argument: "A candy manufacturer many years ago told me that there is little difference in ingredients and products between a 10-pound and a 100-pound batch of candy. But when you get to making a 1,000-pound batch of candy, then there's a difference." Mr. Daniel believes that the popular wine producers and the premium wine makers are simply doing a different job. "There will always be a place for the small operation," he says. "This is where more attention can be given to variety and wines are allowed to age longer and in casks. Take a look at the French cooperative as contrasted with estate bottling. It is the same here: over here you just don't find the Cabernet Sauvignons, Pinot Noirs and Chardonnays in quantity production. There's no ageing in oak casks and no attempt to age in the bottle either."

There seems to be room for both viewpoints. The industrial wines of America may, in time, represent an indigenous American ordinary wine akin to French *gros rouge*. Ordinary wines will always be made and sold in the world, and will usually be made in great bulk, but any wine-growing tradition is likely to lead also to the making of fine wines, even great wines. There seems to be no reason, so far, to believe that America will be an exception to this state of affairs.

WINE FESTIVAL IN MENDOZA, ARGENTINA. *(Photo Leonard McCombe)*

LATIN AMERICA

Anyone interested in grape harvesting or wine making south of the equator should not show up in September or October, since here the seasons are reversed. Vines are pruned in June and August; plowing, sulphur spraying and other treatments take place from November to January; harvesting is done in February and March.

Argentina is, by far, not only the largest producer in Latin America but in the Western Hemisphere, with an annual harvest of around half a billion gallons. The country is thus the world's fourth largest producer of wines, after France, Italy, and Spain.

ARGENTINA

In 1561, in the western part of Argentina, Captain Don Pedro de Castillo founded the city of Mendoza, today the country's wine capital. The following year Captain Juan Jupe founded the city of San Juan, nearby, at the foot of the Andes. As a sort of dowry, he gave the city its first vine plants, brought over from Europe by way of Peru and Chile. Grown first by Franciscan and Jesuit brothers, these grapes were called *criollos* (Creoles), signifying for vine plants, as for people, of Spanish origin but born in the new land. They gave a start to wine growing in the Cuyo region near Mendoza, a region which is today the site of one of the world's largest mass production vineyards.

In 1892, 200,000 plants from the Old World were imported, and an agricultural school was organized by winegrowers, some of them of French origin. Developing the Argentine vineyards was a demanding task, for nearly everywhere they have had to be irrigated. With their vines held up on wires or on *parrales* (trellises) of carob, poplar, or eucalyptus wood, these vineyards present the picture of a Mediterranean landscape, but on an American scale. Grapes grow on a vast, seemingly endless plain, completely flat and uniform.

The harvest, as in any wine country, is gay and animated. The foreman, here called a *corregidor,* rides around on horseback. When trucks drive by to gather the harvested grapes, the *fichero,* who is the owner's representative, throws to each harvester a *ficha,* a token made of copper or cardboard, which represents the future payment for the load.

The grapes go into huge crushers and presses which are capable of pressing

AN AMERICAN-TYPE PRESS in Chile. *(Photo Pan American Union)*

out 20,000 gallons of grape juice per hour, and the wine is stored in glass-lined cement vats that hold up to 50,000 gallons each. The harvest ends with numerous festivities—notably in Mendoza, where they begin on March 24.

The wine best known abroad is El Trapicho red wine, produced on 500 acres of vineyards in Mendoza. Other wines, full-bodied, white or red, are subdivided into ordinary and fine wines. The latter are presented under names borrowed from Europe: Sauternes, Médoc, Saint-Emilion, Valdepeñas, Chianti, and so forth, but the resemblance with their namesakes is rare, and the result of chance.

CHILE

After the Chileans won their independence in 1818, some of them, visiting Europe, were struck by the similarity between the Rheingau and Bordeaux regions and certain valleys in their own country, such as those of Aconcagua, Lontué and, more particularly, Maipo, south of Santiago. Back home the Chileans took the wealth coming from their copper mines and put it to work, importing vine plants and machinery, as well as oenologists, mostly from France.

Most of the Chilean vineyards today, and the best ones, are in the fertile central valley, in the six provinces of Valparaiso, Santiago, Colchagua, Curicó, Talca, and O'Higgins. They cover more than 80,000 acres, but there is a contrast here with the huge Argentine vineyards. Chilean estates (or *fundos*) seldom

VINEYARDS IN THE ANDES, Chile. *(Photo Pan American Union)*

exceed a few hundred acres, and smaller parcels *(quintas)* in the suburbs of Santiago generally run about four or five acres. In the *quintas* oranges, cereals, pastureland, olive trees, avocado bushes, and sunflowers share the land with vine trellises.

The northern vineyards, at Tarapacá, Atacama, Coquimbo, and Aconcagua, produce on some 8,000 acres a number of liquorous and sweet wines from various Muscat-type grapes. After the juice for these wines has stopped flowing from the weight of the grapes themselves, the grapes are pressed to make a wine used for producing the local brandy, *pisco.*

Everyday wines in the central valley are made with Cabernet, Verdot, Cot, Merlot, Folle Blanche (called *Loca Blanca*), Sauvignon and Sémillon Blanc grape varieties. These are, by and large, fair or good. The largest exporter of bottled wines is Undurraga, a winehouse founded four generations ago, and one which has won many prizes in international exhibits. Generally, the label has the seal of the producing winehouse, but the vine plant is not always mentioned.

Among the best bottled wines are a Riesling, called *Grand Vin du Rhin Blanc* and sold in a squat sort of Bocksbeutel bottle, and a red *Grand Vin Pinot Noir.* There are several pleasant, harmless rosés, such as the Chacoli, and a popular sparkling wine (on the sweet side), the Valdivisso. The best wines are labeled *Reservado* or *Gran Vino,* and are subject to government control.

South of Talca, some 150,000 acres of vineyards are almost exclusively devoted to the Raïs, a Creole plant which gives a strong, somewhat wild wine that must be blended to yield the cheap red and white wine reserved for local use. Chilean consumption of wine is quite high, averaging some 15 gallons per head per year. One of the favorite beverages during the harvest season is the *chicha,* or fermenting grape juice.

BRAZIL

Brazilian agriculture received a strong impetus from an influx of Italian immigrants who, from 1875 on, started uprooting centuries-old araucaria pine trees, and opening up clearings and pathways into the forest. The period was that of Pedro the Second, emperor of independent Brazil, who ascended the throne at the age of six in 1831 and reigned during half a century of economic, educational, artistic and agricultural expansion. Immigrants were attracted in great numbers, settling chiefly in the state of Rio Grande do Sul, bordering northern Uruguay and the Atlantic Ocean. There they started vineyards (planted almost exclusively with *labrusca* varieties of grapes) and prospered so well that by 1890 the first Brazilian wine fair took place in the city of Caxias do Sul.

Today some twenty districts derive their livelihood chiefly from wine, and the magnificent vineyards of Rio Grande do Sul, thick and green, checkering the land between patches of trees and red-roofed houses, are reminiscent of a well-kept European wine-growing region.

Many wines are still made from American vines, such as Isabella or Concord, or from hybrids, such as the Herbemont, Jacquez, and Seibels. The wine is generally mediocre, in spite of persistent efforts of oenologists and experimental wine-growing stations, the first of which was founded in Caxias do Sul by Louis Esquier, a Frenchman, in 1930.

Several growers, however, are now producing wine with European plants, and there is a concerted effort to grow these plants in various regions, in order to find out where each one is best suited. In a single vineyard, that of Louis Mandelli, president of the *Sociedade Vinicola Riograndese Ltda.,* 94 types of plants have been experimentally planted.

The best Brazilian wines are sold under the label of the Granja Uniao, and the varietal name is usually given. Cabernets, Merlots, and *Grand Rouge* wines with 11.5 to 12 percent alcohol, are dry and supple, sometimes a bit rough. The Riesling plant has acclimated well in Brazil, producing wines with the pronounced flavor of the grape.

Wines for home consumption are still made chiefly with hybrid plants. The region of Jundiai, in the state of São Paulo, produces up to two million gallons, chiefly to quench thirsts in São Paulo City and Rio de Janeiro.

A new wine-growing district has been started on the left bank of the San Francisco river, in the Nordeste region. European plants can apparently grow here without being grafted on American root stock to protect them from the phylloxera, and the vegetative cycle is so rapid that two harvests can sometimes be made in a year.

Brazilians, who love carnivals, have a number of wine festivals. They are usually noisy and colorful, with balls, outdoor religious services, parades, re-

WINE MERCHANT in regional costume, Brazil.
(Photo Manchette)

CAPPING BOTTLES in
a Parras wine cellar, Mexico.
(Photo Norman Thomas)

ceptions, carnivals, singing festivals and competitions, sporting and acrobatic demonstrations, and the election of the inevitable harvest queen.

URUGUAY

At any time of day or night, Uruguayans seem to be drinking *maté*, an aromatic beverage made from special tea leaves. But when it comes to settling down for a serious meal, such as *asado* (a side of lamb roasted on a spit) or the *asado con cuero* (a whole lamb cooked with skin and hair), they are wine drinkers too.

Endowed with a temperate climate and watered with abundant rain, Uruguay has several wine-growing regions, the principal one being Canelones. It was here, one bright morning in 1520, that a sailor in Magellan's ship sighted the tall El Cerro hill and shouted gleefully "Monte vid'eu" ("I see a mountain"), giving a name to a capital that did not yet exist.

Some of the wines are excellent, particularly those made with European plants such as the Alicante, Carignan, Grenache and Cinsault. The best are usually called *Gran Reserva*, with the added *Tinto* for reds, *Rosada* for rosés, and *Blanco* for whites.

PERU

Peru's vineyards are the world's highest, and its wine growers, the most acrobatic. Grapes often grow at altitudes above 6,000 feet, sometimes overhanging torrents and ravines. One finds them chiefly around Chincha, Lima and Ica.

The first vine plants were planted by priests and monks, after being carried up on muleback along narrow and slippery paths. The wines are not exported but visitors can rely on finding plain table varieties at quite reasonable prices.

MEXICO

Having captured his host, the Aztec Moctezuma, the Spanish conqueror Hernando Cortes organized Mexico into a sort of protectorate, giving to each of his compatriots the responsibility for a certain number of Indians. On March 20, 1524, he decreed that during the following five years, each conquistador should plant one thousand vine plants every year, failing which he would lose his contingent of Indian laborers. This, according to Alberto Jardí Porres, president of the National Association of Viticulturists and Viniculturists, is enough to explain the genesis of Mexican and other vineyards on the American continent.

Local plants were tried at first, but without success. Then came the famous Mission vine, the grape with which Father Junipero Serra evangelized California in the 18th century, and a grape still used for Mexican wine making.

The first vineyard, laid out in the Hacienda de Santa Maria de las Perras (St. Mary of the Vineyards) by Captain Francisco de Urdinola in the 16th century, still exists. Today, throughout Mexico, vineyards cover a considerable acreage, and are still growing. But Mexicans drink little wine—about half a gallon a year per person—and winegrowers are trying to improve their wines and increase their variety in the hope of persuading their compatriots to take some wine instead of their beloved *tequila,* distilled from cactus.

SOUTH AFRICA

Wine has been grown in South Africa for more than three centuries—ever since the first Dutch settlers planted vines in the gardens of the Dutch East India Company in 1655. Jan van Riebeck, head of the first settlement, and also the first winegrower, wrote in his diary on February 2, 1659: "Today, praised be the Lord, wine was made for the first time from Cape Grapes." Two decades later, Governor Simon van der Stel set out vineyards in the Groot Constantia area—still the most famous wine-growing region of South Africa. But the largest impetus came in 1688 when French Huguenots, threatened with religious persecution in France, sailed to South Africa, bringing vine plants from their native Bordeaux or Burgundy. Many of them settled the Franschhoek region, and gave their property French names such as La Provence, Bien Donne, or La Gratitude.

The country lends itself well to wine growing, combining a climate neither too cold nor too hot, neither too dry nor too wet, with soil neither too rich nor too poor. The French immigrants had the knowledge and enthusiasm needed to make much good wine—and they did.

During the 19th century Cape wines were well known throughout Europe, and particularly in England, where fortified Sherry- or Port-type wines from South Africa competed with Spanish and Portugese imports. South African efforts were given a further boost when the British took possession of the Cape in 1806, and when England, faced with a long war against Napoleon, was threatened with being deprived of Claret, a well-nigh indispensable commodity.

South African growers prospered, and much was invested in vineyards and wineries. But then came Waterloo, and true Claret came back to the English market. The South African industry had huge unsold stocks of wine and prices plummeted. The *coup de grâce* fell in 1861, when Gladstone virtually abolished wine duties. As a result, the preferential tariffs accorded to South Africa could no longer stem the flow of French, German, Portuguese, and Spanish wines into England. The phylloxera found Cape winegrowers down on their knees, facing total ruin.

Undaunted, the growers grafted vines onto American root stock, and soon the industry was flourishing again—so much that during World War I overproduction caused prices to fall to a new low. In 1918, wine alcohol was even used as fuel in South African stoves. About that time the growers united to form the Co-operative Wine Growers Association of South Africa, Ltd., or *Ko-Operative Wijnbouwers Vereniging van Zuid-Afrika, Beperkt*, better known as the K.W.V. Preferential tariffs were later restored by the Empire Marketing Board and the Ottawa Agreements, and Cape growers were back in business once more. They prospered in spite of some difficulties encountered by the K.W.V. in maintaining minimum prices, regulating the sale of wine, and handling surpluses. The story repeated itself after World War II, during which many of the European vineyards were destroyed, with the result that South African wine, no longer in surplus, could not fill the demand. Since then the market has become stable once more. Local consumption (about three gallons per person a year) has shown a slight increase, resulting chiefly from improved purchasing power among colored South Africans, who perform nearly all of the manual labor, including grape harvesting.

Practically all of the 150,000 acres of vineyards are in the Cape Province area —and the best wines come from the coastal belt, around Stellenbosch, Paarl, Wellington, Malmesbury, and the famed Constantia Valley. The state now owns the Groot Constantia vineyard, whose wines once reached the cellars of Napoleon, Bismarck, and King Louis Philippe of France.

Red table wines are known either as Claret-type or Burgundy-type, the former being lighter and better. Most are sold under the name of the grape that produces them, accompanied by the name of the area, and sometimes of a "Château" vineyard. White table wines, with a few exceptions, are not as good as reds. There is much sun in South Africa, and therefore the white wines have a high sugar content, too little volatile acids, and tend to taste flat. Many wines, however, are good plain everyday types, and, being relatively inexpensive, can be excellent buys—particularly in Britain. Wine-making methods are modern, and the wine is natural. No sugar can be added, and wines are sweetened either with unfermented sweet must, or with a sweet wine called Jerepigo, prepared especially for that purpose.

The Cape's chief products, rather than its table wines, are its Sherry-type wines—made the same way they are in Spain. Since 1933, South African oenologists have developed their own *flor* bacteria, and no longer use Spanish *flor*. The dry Sherry-type is considered the best and, by and large, South African

SOUTH AFRICAN GRAPE HARVEST

sherries represent an excellent value for someone not overly concerned with the exact quality of a true Spanish one.

There are many sweet dessert wines. These include fair Port-type wines, matured in oak casks, either the full-bodied vigorous and dark "Rubies" or the lighter, more subtle "Tawnies." Tons of Muscatel wine (here the grape is called *Hanepoot*) are made, and much of it is excruciatingly sweet, strongly fortified, cheap, and sold chiefly to the natives.

The purchaser of exported South African wine is almost sure to have a fair wine at a fair price, for export standards, so important to the industry, are strictly enforced. The visitor to South Africa may taste a wider variety but probably nothing better than the exported wines he can buy at home. Depending upon the license of the outlet, wine is sold by the glass (the tot, to be drunk on the spot) or in bottle, to be served with meals, notably with the famed national dish, the *biltong*. This is made of meat—mutton, beef, antelope, kudu, impala—rubbed with a mixture of salt, brown sugar and saltpeter, and hung to dry.

AUSTRALIA

The grape is such a pillar of the Australian economy that Australians have been bombarded with clamorous publicity claiming that their wines are as good as any in the world, if not better. However the Australians themselves don't seem to have been persuaded. The per capita wine consumption wavers between a gallon and a gallon-and-a-half of wine a year. Nevertheless Australian wines are constantly improving. They are also plentiful, and, if the export market holds up, may have a bright future.

Wine-bearing vines were planted on the continent's southern shore in 1788 by Captain Arthur Phillip, the first governor of His Majesty's colony of New South Wales. The colonists had other problems to contend with, however, and Captain Phillip's vines, though they yielded fair grapes, did not give rise to a wine industry. The governor persisted and planted again—at Parramatta, 15 miles inland, where both climate and soil were more suitable. Ten years later the Crown was informed that wine had been produced in Australia—and received as well a request for technical help. Britain was at war with France, and her contribution to the colony's viticultural effort was modest. She shipped to Australia two

SAMPLING WINE in an Australian cellar.
(Photo Wolfgang Sievers)

Opposite page:
TALL FERMENTATION VATS in Australia.
(Photo Wolfgang Sievers)

French prisoners of war who, it was said, were expert winegrowers. Whether from incompetence or patriotism, the Frenchmen failed miserably.

The next effort came from an explorer, Gregory Blaxland, who set out a vineyard along the Parramatta River, and in 1822 proudly shipped to London a quarter pipe (about 30 gallons) of red wine—an accomplishment for which he was awarded a silver medal by the Society for the Encouragement of the Arts, Manufactures, and Commerce. And in 1827, a pioneer of the Merino sheep industry, John MacArthur, produced the first commercial vintage of 20,000 gallons.

The greatest contributor to the Australian wine industry was a Scotsman, James Busby, who came to the new colony in 1824 to run an orphans' school. Busby was a quiet man, but no less effective than his fiery counterpart in Californian viticulture, the Hungarian Colonel Agoston Haraszthy. Busby started teaching his schoolchildren to make wine, hoping this might prove useful for them in earning a living. Then, like Haraszthy, Busby traveled through the wine areas of Europe, sipping wines, improving his knowledge of the art of wine making, and selecting 20,000 cuttings. Most of these he planted in the Hunter River valley north of Sydney, today one of the most important wine-growing districts.

In 1901, the Australian colonies were organized as states within a federal system. Viticulture spread rapidly, first around Sydney in New South Wales, but also north to the Roma district of Queensland, and west to the Barrossa Valley of South Australia and to around Perth, in Western Australia. Until the 1920's, nearly all the wine was consumed domestically. But as production increased, demand lagged, and the government encouraged exportation. The Wine Export Act of 1924 offered a four-shilling premium for each exported gallon of wine. There was further encouragement when Britain adopted favorable tariff rates for Commonwealth products.

In recent years the wine-growing techniques and the quality of the wine have improved. The Roseworthy Agricultural College near Adelaide, South Australia, has a course in oenology so highly reputed that it attracts foreign students, and the Australian Wine Research Institute, founded in 1955 with headquarters in Adelaide, has a staff of researchers and experts who try to improve the quality of the wine—notably, by selecting the plants most suitable for each wine-growing region. The most current plants are Shiraz, Grenache, Muscat, Doradillo, Pedro Ximenes, Palomino and Riesling.

The wines of Australia are often given European place names, although even Australians agree that any resemblance between a local Sauternes and one from Bordeaux is a rare accident. "It is not possible, nor would true wine lovers wish it, to compare Australian wine with the wines of other countries, any more than you can compare the waratah flower of New South Wales with an English rose," states the Australian Wine Board. True enough—but then, if you're not going to call a waratah a rose, why call a Coonawarra Cabernet a "Bordeaux"?

Still, some of the Australian wines can be quite enjoyable, and their popularity in Britain—and, lately, in Japan—has other than financial reasons. The "Sherries" are particularly good, especially the dry *flor* sherries, and so are the "Port-types" (the latter are called Port-types under a treaty signed with Portugal). Many Australian wines are bottled in England.

As for table wines, some of the best are labeled with the names of Australian wineries, combined with that of a region and of a grape variety and sometimes a brand name. Typical examples are Reynella Cabernet Sauvignon or Wynn's Coonawara Cabernet for red wines, and Smith's Yalumba Carte d'Or or Lindeman's Coolalta for whites. There is an Australian Wine Center in Soho, London, where the curious can sample a large selection of wines and brandies.

AN ENGLISH CELLAR MASTER takes his tea in his vineyard. Although very rare, a few true vineyards do exist in England. *(Photo Chapman)*

Wine Appreciation

INDIAN MINIATURE. *Bibliothèque nationale, Paris. (Photo Etienne Hubert)*

Preceding page: FRENCH SILVER WINE-TASTING CUPS, 18th century.
Left to right, top to bottom:
Loire Valley, Ile de France, Poitou, Lyonnais, Burgundy. *(Photo Guillemot Réalités)*

NOW, ABOUT DRINKING IT...

Drink what pleases you with the dishes you like. This is the cardinal—indeed, perhaps the only—"rule" of wine drinking. Wines—or most of them, at any rate —are not a subject to be approached with awe. For the sceptical, some examples may be convincing.

The owner of a renowned vineyard in Burgundy, that had been handed down from father to son for five centuries, once braved the opinions of his companions at dinner by drinking a red Beaujolais with herring. A recent surgical operation had forbidden him the use of white wine, but as he was traveling in Scandinavia, he did not want to deprive himself of the excellent local seafood specialities unavailable at home. And while he violated one of the most cherished rules of the rule-makers—that fish calls for white wine—he had a treat and enjoyed his meal immensely. Another wine lover, a vineyard owner in Bordeaux, was once asked how to drink the prestigious wine of his area. "Open your mouth," he replied, "and pour it down." Both men were expert wine tasters, but neither saw any point in being stuffy about enjoying his favorite drink.

Most wines are best drunk with a meal. The meal is enhanced by the wine and the wine by the meal. This association of wine and food has given rise to certain traditions, most of which are eminently sensible, reflecting as they do the likes and dislikes of a great number of people over a long period of time. It may even be that some tastes are inherent in human nature. Most palates, for example, would balk at drinking a sweet wine with a salami sandwich, as they would revolt when presented with, say, a strong cheese and strawberry jam.

Such basic traditions have been subjected to evolution, which has given rise to the art of cooking food by combining one element with another, be it turkey and cranberry sauce or goose liver and truffles. This also holds true for wines: certain foods combine better with some wines than with others. It has turned out, for example, that most people prefer a dry white wine with fish, and a full-bodied red one with game.

This evolution can be pushed to extremes, and the principles can become more complicated. Herein, however, lies a danger. In becoming more complicated, wine drinking can become more interesting, but it can also—in certain hands— degenerate into a fussy pedantry where the drinker feels so surrounded by rules telling him what he can and cannot do that he hardly has time to enjoy the wine. And wine is to be enjoyed. It can be quaffed wholesale. It can be sipped thoughtfully. It can accompany a meal or a snack. Or it can be tasted with all the dispassion of a professional wine taster. It is up to each drinker of wine to decide what pleases him most.

For many people, as a meal becomes as occasion, a more distinguished wine than that used for everyday becomes appropriate. For the enthusiast or the purist, this choice can be demanding and the association between wine and food becomes both more complex and more intellectual.

There is, of course, in this world of enthusiasts no shortage of talk about wine. Wine-tasting orders, associations, fraternities and clubs are born almost every year. A few have had a long life and a past full of history: the *Tastevins de Bourgogne, Sacavins d'Anjou* and *Jurades de Saint-Emilion* are among the best known in France, and they have their counterparts in many wine-drinking countries. Others are much less publicized—such as the Physicians' Wine Appreciation Society in the United States, with headquarters in New York and chapters in several other cities, whose members gather frequently to examine the fruit of the vine.

Wine-tasting groups meet with varying regularity, partake of choice meals and taste selected wines, sometimes wear traditional costumes in a medieval atmosphere, and usually exchange information about little-known wines and well-furnished cellars. Growers boast of their vineyards, gourmets give recipes, but only seldom is there anyone who drinks too much. For the main rule of wine appreciation is to avoid excess, and instead to combine wines, food, and conversation into a harmonious whole. The approach becomes more intellectual, and the experience more deeply satisfying, because it is a combination of several experiences and memories, just as a symphony is a combination of several well-chosen types of sounds.

"As soon as the meal has started, the genie of the bottle will go from mouth to mouth to give the password: Happiness . . . From the first glass on, tongues loosen and friendships take shape. Wine propagates blissful radiation, insinuates itself into one's limbs, gives courage to the timid, words to the silent, and spirit to all." So speaks Louis Orizet, the Inspector General of the French Institute of Controlled Place Names, and an enthusiastic and highly professional wine taster. He adds: "To compare one's sensations is to intellectualize the tasting of a great wine which, otherwise, would slip toward the prosaic of pure sensualism. Thus not only does wine bring close together the senses and the spirit in man, but it continues to explain away the apparent contradictions between everybody's sensual perceptions, to invite all to commune with the world."

Each one is on his own to judge the value of any particular bottle. Yet many people shun wine tasting, perhaps because they have heard too much about purists and fear they cannot come up to the mark. "The handing over of the wine list to one's neighbor becomes more than just a trusting gesture," comments Monsieur Orizet. "It represents an evasion of responsibility, a sort of inhibition or paralysis of a man's possibility of making a choice—in short, a complete distrust in one's own judgment.

"The more courageous admit their shortcomings by saying it will be enough for them to like—or to dislike—a given wine. Craftier ones hide in sententious silence, less injurious to their pride. There is no merit, in such cases, for the 'expert' to make a choice, to praise or to criticize the wine that is a bond between us. In fact, such attitudes are not limited to wine tasting. Everything that speaks to the senses—music, or painting—may trigger similar reactions. Such a phenomenon deserves some thought: If most people doubt their judgment, doesn't it mean that they suspect, more or less consciously, that it is fragile?"

A DEAN of French wine stewards. *(Photo Boubat-Réalités)*

THE THREE WINE SENSES

To choose a wine, to appreciate it, is to make a judgment. We all start with roughly the same equipment, the same potential for wine tasting: our senses and the understanding of the messages they carry.

The first sense to be brought into play is sight: Is the wine a thick, troubled broth, hiding behind a doubtful veil? Or is it a brilliant, appetizing liquid, with a golden yellow, amber, straw color? Or a color that looks tarnished, milky and artificial? Does it have a cheerful open-minded red, ruby, garnet or crimson color, or a discouraging and gloomy darkness?

Shake the glass, and a bit of froth appears: if it is white, it bespeaks an old wine. If red, it may mean the wine is young and has some bubbles, or that it is "plastered" (has had too much calcium sulphate or gypsum added to clear it and to help it age) or acid. Exceptionally, the froth may come from a few particular varieties of grapes that are known to be rich in pigment. If all goes well, this froth or foam rapidly vanishes. If it does not, the wine betrays its weakness in alcohol.

The second and most important sense is smell. This identifies the odoriferous and volatile molecules which are exhaled by wine and are forced upward by the warmth of the mouth and movements of the tongue, by breathing, and by swallowing. The result is what most people call taste.

Much can be told about a wine from the way it smells. A young wine will exhale the scent of the grape. It has, also, the fragrance of the yeast that carried out the fermentation. During the summer after it has been harvested, it starts acquiring its bouquet. The bouquet should be straightforward and well balanced: the wine should not have too much volatile acidity, which has a vinegary smell. There should be no smell of sulphur, or of sediment, of cask, of wood, of mold, of rancidness, or of old vat. Even at this stage the wine can be identified as common if it smells of grass, or as fine if its aroma is that of grape.

"WHATEVER HAPPINESS the hand cannot seize is only a dream." A few examples of stemmed glasses, the only kind adequate for fine wines. Left to right: for red Bordeaux, white Bordeaux, Champagne, Burgundy, Rhine wine, and, on the far right, a "flute" and a "coupe" for Champagne. *(Photo Chapman)*

It is only after the bottling that wine develops all of its bouquet, which slowly becomes subtle and, at the same time, more intense. Thus there appear some of the perfumes which Louis Orizet and his wine-tasting friend, Jules Chauvet, discovered in various Beaujolais: rose, jasmine, hyacinth, lilac, orange flower, violet, mignonette; pepper, sandalwood, clove; vanilla, incense; apple, bitter almond, anise, raspberry, and English candy; musk, seaweed, amber, mushroom, venison; liquorice, resin and tobacco.

Compared to the paramount part played by the sense of smell, the contribution of the third sense, taste, as identified by the tongue, seems to be limited. There are, in fact, only four different basic tastes. At the tip, the tongue perceives sweetness. Along a rather large area of its surface on top, it perceives saltiness. On the edge, so to speak, it feels acidity and, finally, the rear part, reached when the object that has crossed the lips is about to be swallowed, perceives bitterness. If one attempts to identify them separately, these various sensations seem to succeed each other, and bitterness can follow sweetness or acidity by as much as ten seconds.

Thus the wine taster can distinguish three stages in his customary operation: the stage of the "attack," which lasts but two or three brief seconds; the evolution of the taste, during which he inhales and rolls the liquid on his tongue, runs it between his teeth and along his palate, an operation accompanied by a suggestive gargling noise; and, finally, the aftertaste, which comes after the liquid has been swallowed—or spit out. The gargling noise comes as the taster draws air into his mouth, through the wine, then exhales the air through his nose. (The neophyte may find he concentrates so hard on the mechanics of this procedure that he forgets to register his sensory impressions.)

The feeling of warmth triggered by alcohol does not belong in the gustatory domain, but in the tactile realm. It is felt by the palate, which recognizes warmth with the inner surfaces of the cheeks and underneath the tongue. This feeling can be translated in various ways. Louis Orizet sees there a notion of the "shape" of the wine: the blossoming of a great wine in the mouth gives a feeling of wildness, sometimes expressed by the saying that a wine "spreads out its peacock tail." The vertical dimension is soaring toward heights, brought about by the emotion of an exquisite discovery, just as the opposite direction—depth—results from a delicious dive into a glass of great Burgundy.

Edouard Kressman, wine merchant in Bordeaux and poet in his spare time, noted judiciously that the only sense not activated by wine tasting is hearing. "And this is why," he concluded with irreproachable logic, "wine tasting and music are complementary arts."

The point is to know how to profit from the sensorial team at your disposal. Wine tasting is not reserved to professionals. It should not be forgotten that the professionals go through their many exacting analyses for the consumer's benefit. Their judgment may be severe, but this is because they cannot afford to make an error that will mislead the buyer. The consumer need not try to take the expert's place and try to figure out every component that makes a wine taste just so, but he can decide whether or not a wine suits him, and why it does, or does not.

Godfrey de Bruyn, an enlightened South African wine lover, suggests a simple procedure for appraising wines: "This is an interesting and pleasant experience, requiring no special technique or special effort. They (amateur wine tasters) need but a bucket, a jug of water, a few sheets of paper and pencils, glasses, a bit of cheese, wines to be tasted, and friends likely to be interested in wine. The choice of wines for a wine-tasting session is a matter of preference, but it is a good politic to limit one wine tasting to one type of wine—such as, for instance, Sherry, Port, or white, dry table wines."

An experienced professional wine taster can arrive at a definition of a wine in fifteen minutes and express it in thirty words, and can repeat the same performance thirty times in a row. Such an ability requires daily training for at least a year, a training which may be lost—or at least impaired—after as little as a week of idleness. Skill of this kind is not expected of everyone, but everyone can express a valid personal opinion, whose value depends on his native sensitivity and his attention.

THE WINE VOCABULARY

People who are thoroughly lacking in such sensitivity, or who are completely absent-minded, are rare. But there are many individual variations, which may account for the preference for one type of wine over another. For instance, some people can detect the presence of a one-thousandth part of sugar in water but perceive bitterness only when it is overwhelming. Others, however, cannot detect ten times as much sugar but balk at the slightest bitterness. Some will cringe at the hint of acidity, while their neighbors will drink, without batting an eye, a wine that begins to turn into vinegar. One must take stock of one's gifts and shortcomings, and then train oneself to recognize the participation, however in-

A BASKET OF GLASSES. Painting by Stoskopff.
Musée de l'oeuvre de Notre-Dame, Strasbourg. (Photo Giraudon)

direct, of different components in a wine. At every new tasting, the wine lover tries to describe what he sees, what he smells, what he tastes, with the most suitable words he can find. This gives birth to that much-maligned wine vocabulary, which makes a neophyte shrug in discouragement or smile with irony. Once he gets used to it, however, a few dozen adjectives become, rather than cumbersome, a simplified code, and a means of communication.

An expert's evaluation is based on the concept of harmony. A good wine implies a judicious balance, chiefly between two pairs of elements: alcohol and sugar on one side of the scale, acid and tannin on the other. One of the foremost authorities in the matter is Professor Peynaud, Bordeaux oenologist, who has put down, to help wine merchants, cellar masters, and wine tasters, two tables of gradation.

First, wine is graded according to its alcohol content. With a low alcohol content, a well-balanced wine can be fresh; if it is poor, it can be small, weak, flat, washed out, or watery. Then (usually from 12 percent alcohol and up) a wine is vinous, warm, generous, full of spirit (and, if unbalanced, simply too strong, or alcoholish).

Secondly, wine is graded with regard to its wealth of components (or dry extract, meaning what remains when wine is boiled away). A well-balanced wine can then be qualified as fluent, velvety, tender, supple, round, mature, full or fleshy. A poorly-balanced one can be skinny, biting, crude, hard, acid, green—these being steps of increasing acidity.

With regard to volatile acids, a wine may appear warm, or "upgoing." It can

VENETIAN WINE GLASSES, 17th century. *Beurkeley collection, Berlin. (Photo Fiorentini)*

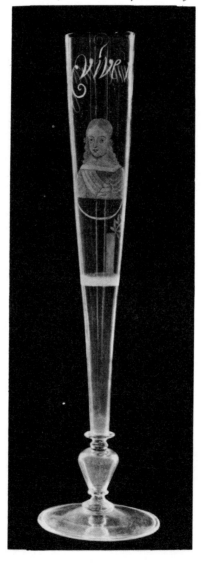

be vinegary, acrid, and even burning, if it has, at the same time, too much alcohol. Finally, excess of bitterness is said to make wine rough, hard, astringent ("puckery"), and bitter. This is a prelude to an examination that could go on *ad infinitum.* Not only does the vocabulary vary with the experience and the imagination of the taster, but subjectivity is such that an eloquent host can lead a drinker to say just about anything about a wine.

The message of the senses will also depend on the order in which the different wines are presented. Accumulation of alcohol, which penetrates in small amounts into the mucous membranes—even if the wine isn't swallowed—will progressively dull the senses. Habituation to a well-known wine makes the drinker less sensitive to its characteristics.

TIME IS OF THE ESSENCE

To bring all these contingencies into consideration, a wine lover should, first of all, take time. A symphony, to be appreciated, cannot be rushed; rather, it should be listened to a second time. Likewise, to rush a wine will risk missing its harmony. The time of day can influence wine tasting. Professional wine tasters usually prefer to taste wines in midmorning, before lunch, when the taste buds are rested. For the same reason they often skip breakfast; and many of them avoid smoking so as not to affect their sense of smell. During a meal, the sense of taste becomes somewhat dulled. The wine does not stand on its own, but is part of a combination. This may be a good time to enjoy drinking wine, but not the best for an objective wine-tasting experience. As to the digestion which follows, it too detracts from a drinker's sensitivity.

It is interesting to taste wines that are not too familiar—but not to try too many in a row: they will appear increasingly harsh and acid. A red wine after a white one seems stronger in tannin, and a dry wine after a sweeter one seems to be even drier and more acid. The environment is also important. Wine-tasting rooms in large winehouses are as neutral as possible, resembling a laboratory, with white walls, and fine glasses of average size, all with a stem to hold them by so that the hand does not unduly warm the wine. Glasses are plump and rounded, so that the liquid can be rolled in them, spreading it around the edges to increase the surface which exhales the bouquet. For the same reason, most wineglasses are tulip-shaped rather than cylindrical, with their opening narrower than their belly, in order to concentrate the aroma exhaled by the wine.

Monsieur Peynaud advises that if several wines are tasted each should be graded from 0 to 20—from the undrinkable to the supreme, with all the intermediate shades of bad, lower than average, ordinary, good with slight imperfections, or excellent. (Grading from 0 to 20 corresponds to grading in the French schools, and an American wine taster may feel more at home by grading wines from F for failure to A plus for unsurpassed excellence.)

Professional diagnosis, of course, is more exacting. Here are a couple of examples of tastings made in Bordeaux. Professional comments often include the results of laboratory analysis, such as alcoholic content, acidity, or tannin index.

DECANTER AND GLASS with enameled decoration bearing the inscription "M. Melicot, curé de Gésier." French, 1754. *Musée des Arts décoratifs, Paris. (Photo Chapman)*

Clairet
Alcohol: 12.2%. Total acidity: 3.80. Volatile acidity: .042. Tannin: 0.8 grams. Supple. Round without being fleshy. Light in body. Well balanced. Pleasant finale. Has taste. A bit of carbonic gas (shake the bottle; the gas fizzles under the thumb). Very good type of Clairet wine. Well representative. Different from rosé. Average quality equivalent to Bordeaux supérieur. Regional plant: neither Médoc nor Saint-Emilion. Very good grade for a Clairet: 15 out of a possible 20. Origin: a cooperative which has processed the first plants it received: Malbec and Merlot. With Cabernets the wine would have been less fluid, less light, less supple.

Pauillac 1959
Alcohol: 12.6%. Total acidity: 4.21. Volatile acidity: .075. Tannin: 2.28 grams. Potent bouquet, very rich and very complex. Noble tannin. Wine is very mature. More vinous and more high-strung than usual Bordeaux, with a firmer finale. Some excessive volatile acidity, but round, full, powerful, full-bodied, with very good tannin that insures longevity. Volatile acid is a minor shortcoming since wine is very good otherwise. A bit dried up, but not harsh. Will never be more supple.

WINE FOR TONIGHT

Tonight the host may want to have an exceptional bottle on his table. He must first choose it—either from his cellar or from the offerings of a wine merchant, whose advice and comments may evoke a previous experience and help the choice. The wine must then be brought to the right temperature—slightly under room temperature for red wines, slightly chilled for white. It must be uncorked, more or less in advance depending on the wine: several hours before serving for some great Bordeaux wines, which have aged long and must be awakened (by breathing oxygen) from hibernation. On the other hand, plain wines can be opened just before serving. The drinking of a well-chosen wine can contribute to create a warmth, a creative and cheerful atmosphere—and there is no need for the menu to be extracted from the annals of a royal or princely court. A roasted or grilled meat followed by a few good cheeses are wonderful companions to a great red wine. Shellfish, oysters or fresh fish generally call for dry, fruity white wines. White meats, fish in a sauce or creamy desserts can combine with a not too austere dry white wine or even white, liquorous dessert wines. With salads, there should be no wine at all, for vinegar or lemon would damage any wine. These basic principles should be kept in mind, which doesn't mean they cannot be violated—judiciously.

Nor do they mean that wines served during dinner with friends must of necessity be great. They can, instead, be plain. An ordinary Italian Chianti, a Californian Cabernet Sauvignon, a simple Beaujolais can be great companions to a plate of spaghetti or a roast beef. But such wines should never be served stingily (and no wine should be served in thimble-sized glasses). Some wine lovers even like red wines served cool, particularly during summer. Without going so far as dropping an ice cube into the glass, the serving of a plain red wine at cellar temperature, or after placing it for 15 minutes in the refrigerator, should not make anyone but wine snobs frown. (Chilling any poor wine, of course, helps hide its shortcomings.) Likewise, the drinking of a light white wine with Italian *pasta* food, with hamburgers or steak, can be quite pleasant, as can be that of fruity and light red wine with fish, or with an omelet.

Nowadays most people have no cellar, but rely on a liquor store, which has a choice of bottles of various origin and in various price ranges. It is wise, as a rule, to dismiss promptly any overly ornate or gilded labels, folkloric wrappers, and other obvious manifestations of hard-sell advertising techniques. A growth praised by a friend may be selected, or a compromise reached between the memories of one growth and the availability of one from a neighboring area. If the contents are satisfactory a case may be worth ordering, and a corner of the basement set up as a cellar for wines to age—undisturbed and in a horizontal position. In the absence of a cellar, a closet may be used, but a cool, constant temperature is preferable for long ageing. Eventually a wine cellar builds up. Each bottle is familiar and a friend, ready to render service when required.

If perchance the wine lover happens to travel through one of the regions from which a wine he knows has come, it is worth while to visit the producer's or the

THE BLACK CURRANT TART. Painting by
William Claesz Heda. *Musée de l'oeuvre
Notre-Dame, Strasbourg. (Photo Etienne Hubert)*

merchant's cellar, to explore a new wealth usually revealed with friendliness.

Wine then becomes a part of life. A connoisseur's health, mood, family life, and relations with friends can be improved. Wine is usually associated with sobriety: "It is not good wine that makes men drunk; it is men who make themselves drunk," observed Confucius.

CORKSCREWS

The purpose of the corkscrew is to pull corks out of bottles with a minimum of damage to the cork, the wine, and the user.

The important part of the corkscrew is its wire spiral which penetrates into

the cork. It should not crumble the cork, but worm itself into it as gently as possible, and hold it firmly as it is pulled out. It should be long enough (about 2½ inches) to go through the entire cork, so that the point protrudes at the other end; otherwise the risk is that only the dry part of the cork will be pulled out.

One fairly common type of corkscrew has a "worm" consisting of a straight central part with a sharp point, surrounded by a spiral (often with sharp edges) which easily penetrates the cork. This is *not* a corkscrew, but a gimlet, to bore a hole in the cork. It works well only with a short, relatively loose cork, but anything longer and more substantial, such as the cylindrical corks used for fine wines destined to age, is not likely to come out. Instead, the gimlet will come out, leaving a hole in the cork, and the user will have to make a second attempt, drill another hole, finally extract pieces of broken cork with a knife, precipitate debris into the bottle, spoil his wine—and his disposition. Several physicists who have studied the principle of such corkscrews have recommended that they be thrown away.

A true corkscrew has a worm which makes an open spiral. The sharp point of the worm should not deviate from the spiral: a true corkscrew can be easily identified by introducing a toothpick or a wooden match through the spiral. The most effective worm does not have a circular cross-section, but a square one, which has less tendency to be pulled through the cork. This worm can be attached to a handle, a lever, a double screw, an expandable pulley-type device, an electric motor, or anything else. But it is the worm that counts.

A simple, small, handy corkscrew is the one that looks like a knife, with the worm folding against it, and with a lever that fits the edge of the bottle to reduce the required effort and avoid shaking the bottle violently during the operation. Such shaking is almost inevitable if the bottle is held between the knees and if the corkscrew has a simple handle. The knife-type corkscrew is used by nearly all French wine waiters. It usually has a strong blade for cutting the bottle-cap neatly away. An advantage of this corkscrew is that it is one of the least expensive ones, and this perhaps is why it is not easy to come by.

CORKSCREWS, 18th century.
Musée des Arts décoratifs, Paris. (Photo Chapman)

SOME RECIPES

Some people miss out on the pleasures of wine cooking because they think it takes "too much time and money" and that the results are "heavy" on the stomach. Yet once these dishes are on the stove they simmer and cook by themselves . . . or almost (they do have to be checked occasionally to see that they don't burn). Most wine dishes can be warmed over, and all are adaptable to pressure cookers.

Foods cooked with wine, though perhaps not as light as grilled meats, are not indigestible . . . on the contrary. Nor are they expensive. Though "red ink" type wines are out of the question, to use fine vintage wines would also be foolish. An average table wine nearly always fills the bill.

Here we have chosen some typical dishes from France and other countries. In some cases the recipes have been simplified; all have been rigorously tested.

Recipes calling for Champagne and Port should be reserved for special occasions. One wine may always be substituted for another. The dish's flavor will vary, depending on how full-bodied or fruity the wine used, but the end result will invariably be delicious.

One point to remember: Never use a sweet white wine in cooking. It darkens as it cooks.

And now a word about the famous cold "wine soup" that is such a summertime favorite with peasants . . . and others. Mix equal amounts of red wine and sugared water, stir well. Then pour over thin slices of bread or croutons fried in butter (the latter a refinement) and serve chilled. In Touraine this is a *miot*, in Poitou a *migé*, in Deux-Sèvres a *soupine* and in Berry a *midonnée*.

The Bordeaux Region

Sausages and oysters (from the Arcachon or Tremblade beds) are typical of the Gascony and Saintonge regions. Oysters are eaten at the same time as small hot pork sausages which have been grilled and then simmered in white wine.

Truites à la bordelaise (Bordeaux-style Trout). For four to six trout make and cook a court bouillon consisting of a half bottle of red Bordeaux wine (or half wine and half water), onions, carrot, salt, peppercorns, thyme, parsley and a little bay leaf. Reduce heat for half an hour, then add the trout to the court bouillon and let them cook for ten minutes. Remove the trout carefully. Strain the court bouillon, check seasoning, thicken with a lump of butter blended with a little flour. Heat the sauce once more to the desired consistency, reheat the trout in the sauce and serve when it comes to a boil. Serve with boiled potatoes or croutons fried in butter.

Variation: Use a white wine such as a Monbazillac instead of a red one.

Escargots à la bordelaise (Bordeaux-style Snails). The basic preparation of snails is always the same.

1. Make a court bouillon of equal amounts of dry white wine and water, with pepper, thyme, bay leaf, sliced carrots, onions and leek. Let this boil a good hour. Count on two quarts of liquid for four to five dozen fat snails.
2. Wash the snails in cold running water or in several changes of water. (The squeamish may let them fast 48 hours beforehand, though this will not improve them in any way.)
3. Let them stand two or three hours in cold water to which a one-tenth part of vingar has added (in order to coagulate the slime), together with a handful of rock salt. (Ten minutes will suffice it the snails have fasted.)
4. Boil snails in salted water 15 minutes; cool in cold water.
5. Extract the snails from their shells with a nail or strong pin. If the snails are especially large, cut out the black part (which contains the bitter liver) with scissors.
6. Strain the court bouillon and let the shelled snails simmer in it for three hours.

Once the snails have been cooked according to the above instructions, fry them in a hot pan with butter to which garlic, chopped parsley and bread crumbs have been added. Serve very hot.

If your guests are real snail-lovers, count on at least a couple of dozen small gray snails per head.

Civet de lapin de la garenne (Hare or Rabbit Stew). For one hare use 3¼ ounces fatback, 3¼ ounces bacon, 3¼ ounces onion, 8 ounces mushrooms, a lot of thyme, a little pepper, salt and one or two tablespoons flour. Have the hare or rabbit cut in small pieces.

Cube the fatback and bacon, then fry them together in a stewpot. Take out the bacon, brown the hare or rabbit in the fat. Add onions and when they start to brown, sprinkle with flour, salt and seasoning. Add the cooked bacon and mushrooms; cover with red wine. Let simmer covered for an hour to an hour and a half.

Hare is more delicate and flavorful than rabbit but both may be cooked in the same manner. Rabbit requires more seasoning, especially pepper. A pinch of cinnamon and/or nutmeg is not out of order. Always use red wine.

La compote de grande-mère (Grandma's Stewed Fruit). Melt on the stove 8 ounces of granulated sugar in a pint of water and an equal amount of red wine. When the mixture reaches boiling point, add two or three pears and the same number of peaches, peeled and cut into small pieces. When they are cooked put the pears and peaches in a bowl and reduce the syrup. When the desired thickness has been achieved, add red and white grapes. Allow to cook for five minutes, add a pinch or two of cinnamon, check for flavor, add more sugar if necessary and pour over the stewed fruit. Serve chilled.

Variation: Serve with butter-fried croutons or a bowl of cream.

PEARS IN WINE, a Burgundian recipe (see page 297). *(Photo Chapman)*

Le pudding aux prunes (Prune Pudding). Soak a pound of prunes in strong tea (in the Bordeaux region prunes from Agen are particularly favored for this dish). Drain and then allow them to marinate 12 hours in sweetened red wine. Remove pits, cook in this syrup, then pass them through a strainer. When cool add from 1½ to 3 ounces heavy cream. Place in a mold or salad bowl and set in refrigerator for several hours. Unmold and serve with whipped cream.

Variations: Use fresh plums, in which case they need only marinate in the wine for an hour. Or add the juice of one lemon to the wine.

Burgundy

Boeuf bourguignon (Burgundian Beef). For four people use about two and a quarter pounds round or chuck beef, cut into 1½-inch cubes. Brown in a stewpot, using equal parts oil and butter. Drain the meat and remove grease from pan. Melt a chunk of butter and add meat, sprinkle with flour and stir over a high flame. Pour in red wine to cover, add salt, pepper, parsley, a sprig of thyme and a quarter bay leaf. When it comes to a boil, lower the flame and let simmer for at least three hours.

Meanwhile, fry 5 ounces of streaky bacon cut into cubes, 3 ounces small onions and 5 ounces flap mushrooms, or lacking these, ordinary mushrooms or dried ones which have been previously soaked (at least two hours) in water. Half an hour before serving add the bacon and vegetables to the meat. Check seasoning and skim grease from the sauce if necessary. Serve piping hot accompanied by boiled potatoes, buttered noodles or Spanish rice. This dish is excellent reheated and may be prepared a day or two in advance, making it ideal for Sunday dinners.

Variation: Marinate the meat from 12 to 24 hours in red Burgundy wine along with pepper, carrots, parsley and onion. Drain before browning in the oil-butter mixture, then proceed as above. At the very last instant, bind the sauce with two tablespoons of heavy cream or you may add a half cup of good brandy and flambé it.

Coq au vin (Cock in Wine), the pride of many another province, is also excellent when white wine is used, and can be made according to the same recipes as for Boeuf Bourguignon or its variation. However, though the sight of *Coq au vin* on a menu may bring to mind the noisy king of the barnyard, remember it's best to choose a nice plump chicken rather than a tired old rooster.

Variations: Sauté gently in butter a sliced carrot, a sliced onion, a few veal bones and a sprinkling of flour. When all are lightly browned, add a bottle of red wine, salt, pepper, thyme and bay leaf. You now have a *meurette*. Reduce to low flame for an hour. Brown pieces of chicken in equal parts of oil and butter, add salt and pepper. When chicken is cooked, pour the reduced sauce over it. Correct seasoning if needed; let cook another 15 minutes before serving garnished with croutons.

Oeufs en meurette (Eggs in Wine Sauce). Make a meurette as indicated above and poach eggs in it. Place them on bread slices browned in butter and cover with the sauce which has been thickened with an egg yolk and butter beaten with a wire whisk.

Variations: Use a meurette made with Chablis and serve the eggs with mushrooms or onions cooked separately in Chablis. Or make a wine sauce, then poach eggs in water to which a little wine or even vinegar has been added. Drain, place on croutons and cover with sauce.

Poires au vin (Pears in Wine). Choose small firm pears, counting on at least two per person. Peel, leaving stems. Dissolve the equivalent of a demitasse spoon of granulated sugar per pear in a little water, add pears and cover halfway with red or white wine. Bring slowly to a boil and cook for 20 or 30 minutes, depending on the ripeness of the fruit, which when cooked should be translucent and easily pierced with a knitting needle. Place the pears in a serving dish and cover them with the syrup.

Variation: As the pears start cooking add a pinch of ground cinnamon, grated nutmeg or even a clove.

Champagne

Melon au champagne (Melon in Champagne). Champagne brings out the best in a melon, even more so than Port or any other wine. Select small melons (one per person), cutting a plug around the stem and carefully scooping out the seeds with a spoon. Sprinkle with a little sugar and fill with champagne flavored with a few drops of kirsch. Serve ice-cold (after at least an hour in the refrigerator).

A fruit salad sprinkled with champagne (semi-dry or sweet) is also delicious.

Soles au champagne. Place soles in a shallow dish and cover with champagne; let them marinate for about half an hour. Drain the fish, dust lightly with flour, and brown them in butter, adding salt and pepper to taste. Heat the champagne in which the soles have marinated, adding a tablespoon of heavy cream for each fish. Pour this over the fish in the frying pan, bring to a boil and serve very hot.

Filets of sole may be prepared the same way, and *blanc de blanc* wine can serve nicely to substitute for champagne in this dish. Furthermore, veal cutlets cooked in this manner are excellent.

Loire

Rognons au Juliénas (Kidneys in Juliénas). Remove excess fat and membrane from veal kidneys. Split them in half, remove fat and skin, slice thinly. Brown in salted butter, dust with a little flour, add pepper. Moisten the kidneys well with Juliénas wine and let them cook over a low flame for ten minutes or so. Serve with "roesti" potatoes (a kind of potato pancakes made with sautéd onions and browned in butter) and a bottle of the wine you used for the sauce.

Beef kidneys, not as fine but considerably cheaper than veal kidneys, can also be cooked in this manner. However they should be washed first in boiling water to get rid of their ammonia-like smell. After draining them in a colander, steep the kidneys in Juliénas for an hour or two to tenderize them. (The wine used in this process can also be used in cooking them.)

Calf liver—or even beef liver—is superb yet inexpensive cooked in this way.

Cuisses de grenouille au brouilly (Frog Legs in Brouilly). For centuries France was the only country in Europe, if not in the world, where frog legs were eaten. Nowadays many non-French gastronomes have happily acquired the habit. Ranch-grown frogs are available throughout the year, but although plumper they lack the subtle flavor of their swamp-bred counterparts.

Lightly flour the frog legs, sauté them briskly in hot butter. Salt, pepper, cover with Brouilly wine and let them simmer ten minutes. Thicken sauce with egg yolk—or cream—over a very low flame. Do not let the sauce boil, but serve as hot as possible.

Variation: Prepare a white wine sauce and simmer floured frog legs for ten minutes. At the last moment thicken the sauce with egg yolk or cream.

Rhone, Provence, and Midi

First, just a word about Savoy fondues. These are made and eaten in the same manner as those in Switzerland (see recipes under SWITZERLAND) except that a local dry white wine is used.

Les écrevisses en buisson (Crayfish in a Hedge). Buy four to six crayfish for each person. Remove their tails and main fin so that the lower end of their digestive tract may be extracted. Make a strongly seasoned court bouillon, using white wine (and no water). After it has been reduced, place the crayfish in it and let them simmer for ten minutes or until they turn a gorgeous red. Leave them in the court bouillon until they are ready to serve, warm and stacked vertically on a platter to form a hedge.

Les brochettes d'écrevisses (Skewered Crayfish). After the crayfish have been cooked, chop off and shell the tails, then thread them onto a skewer. Dip them in melted butter and grill rapidly. A real feast.

Les oeufs au rosé de Provence (Eggs in Rosé de Provence). This is the same recipe as for Eggs in Wine Sauce (see recipes under BURGUNDY), except that Rosé wine is used.

Daube à la provençale (Provençal Beef Stew). Prepare a marinade with red wine, peppercorns, thyme, bay leaf, estragon, parsley, lemon slices, an onion and two or three garlic cloves. Macerate two and a quarter pounds of beef (rump or flank), cut into one and a half inch cubes, in the marinade for at least 12 hours. Drain the meat, then brown quickly in very hot olive oil. Add the marinade; if it does not cover the meat entirely, add wine—the same kind you used to make the marinade. Let the whole simmer over a low flame for two hours. Skim off the grease but do not strain, and serve hot with a sprinkling of chives.

Chicken treated the same way is marvelous.

Variation: Add to the marinade an orange cut into slices, and the juice of one orange to the sauce during cooking.

La daurade à la provençale (Provençal Red Snapper). Scale, clean and slice a good-sized red snapper (two and a quarter pounds for four people). Fry the slices in olive oil, salt, pepper, parsley, garlic and onion. Cover with red wine and let simmer for an hour. Just before serving, thicken the sauce with an egg yolk or a little cream.

Variation: Use the same method for burbot or carp.

Les saucisses de Toulouse au vin blanc (Toulouse Sausages in White Wine). Grill the sausages in a very hot dry frying pan after first pricking them to prevent bursting. As soon as they have "given up" their fat, pour the fat off and keep to one side. Add a little white dry wine to the sausages, lower the flame and let them simmer for 15 minutes. Serve with potatoes which you have in the meanwhile fried in the sausage fat.

La matelote d'anguille à la béarnaise (Béarnaise Eel Stew). Clean and cut good-sized eels into sections, counting on two sections per guest. Fry the pieces in a stewpan, using a half-oil, half-butter mixture. Flambé with a small glass of cognac. Remove the eel. Brown sliced onions in the oil-butter mixture; add a little flour to make a paste, or roux. Replace the eel in the pan, salt and pepper lightly, and add red wine. Check cooking after half an hour. Serve in a shallow dish garnished with croutons (fried in butter or oil and preferably rubbed with garlic beforehand).

Les framboises du bon roy (The Good King's Raspberries—referring to René I of Provence). Marinate raspberries in sweet white wine, to which sugar has been added, for an hour or so. Drain. Beat eggs into omelette condition—two eggs for every half pound of raspberries. Place drained raspberries in a buttered baking dish, pour the eggs over them and sprinkle with sugar. Let the mixture set in the oven and serve hot.

Variation: The same dessert made with strawberries or grapes.

Melon de la Têt au banyuls (Melon Balls in Banyuls). Cut a melon into cubes or make balls and sprinkle with Banyuls wine. Serve chilled in individual goblets as an appetiser or dessert. Let guests add their own powdered sugar to taste.

Alsace

Le coq au riesling (Cock in Riesling) of Alsace is just as famous as the *coq au vin* of Burgundy or Chanturgues, and is made in the same way except for the wine.

La daube à l'aigre-doux (Sweet and Sour Pot Roast). Buy a two or three pound pot roast and marinate for three or four days in red wine to which has been added an onion, garlic, carrot, parsley, and estragon (all tied together), also pepper and four lumps of sugar. Drain the meat and brown it in oil. Add the marinade, salt, and let the meat stew slowly for two and a half to three hours, depending on the size of the roast. Correct the seasoning, then add a generous handful of raisins which have first been soaked. Let cook another ten minutes. Thicken the sauce with a little cream and serve it in a sauceboat.

Variations: Proceed similarly, using white wine.

Sèche au vin (Wine Shortcake). Line a fairly deep pie tin with pie dough. Prick with a fork and cover with a thin layer of sugar mixed with a pinch of flour and cinnamon (to taste.) Fill with white wine to a depth of ¼ inch. Dot with small lumps of fresh butter and set in a hot oven. Take care not to use too much wine as the sugar-wine mixture unfortunately tends to overflow in the pie tin while cooking. Serve only after it has cooled completely.

Variation: Cook the wine with sugar, cinnamon, flour and butter to obtain a thick syrup. Place half this on the dough before it goes into the oven, adding the remainder in two or three portions as it cooks. The cake will be softer this way.

Vin de framboise (Raspberry Wine). Use two pounds of raspberries for every quart of red or white wine. Crush the fruit and pour the wine over it to marinate for 48 hours. Stir with a mixer or blender and then strain through a sieve or cheesecloth. Add a pound of sugar for every quart of liquid, place over a medium flame and stir constantly so that the sugar dissolves. Boil for four or five minutes. Take from stove and add one quarter pint 180-proof alcohol. Allow this to cool, then bottle and cork. Two months' "ageing" is a minimum.

Franche-Comté

Though detached from Burgundy in the 9th century, Franche-Comté has remained its twin sister insofar as cooking is concerned. Eggs are poached in Arbois wine, *coq au vin* simmers in Arbois, and *truite au bleu,* or boiled trout, are a delight not only for the quality of the fish—which are kept in streamside vivariums—but for the red, rosé or white wines used in the court bouillon.

Truite au bleu (Boiled Trout). Count on roughly a half-pound trout for each person. Trout must be cooked immediately upon its departure from the vivarium —it must be thrown alive into a court bouillon made of either pure wine or a half-wine, half-water mixture seasoned with pepper, salt, parsley, sliced carrots and a small onion. Traditionally this court bouillon is allowed to reduce for at least an hour, but I find the time can be cut to a minimum if it is made stronger.

Serve these trout in a little court bouillon or on a hot plate draped with a towel, accompanied by boiled or steamed potatoes and melted butter.

If the trout is not thrown alive into the pot, it will simply be trout cooked in a court bouillon. In this case, serve it with a sauce made of a white roux moistened with the liquid of the court bouillon.

Italy

La stufata (Italian Pot Roast). Marinate one quarter pound of bacon, cut into cubes, in a pint of Chianti seasoned with thyme, bay leaf, basil, sliced onions, carrots and turnips (3¼ ounces of each of the last three). Lard a 2¼-pound roast beef with the pieces of bacon and brown in a quarter pound of pork fat. Add the vegetables from the marinade and let them simmer. Pour in the marinade and add 3¼ ounces of dried mushrooms which have been soaked for a few hours in water. Cook over a very low flame for about four hours. Serve covered with the strained sauce.

Variations: Marinate the meat for 12 hours in white wine seasoned as above. Or instead of beef, use a veal or pork roast, but without bacon in the marinade or the meat.

Poulet au vermouth (Chicken in Vermouth). Stuff a chicken with a small onion, two peppercorns, a small glass of Turin Vermouth and a lump of butter. Brown the bird in olive oil. Salt. Turn the flame down, sprinkle with a half cup of Vermouth and the same amount of bouillon. Cover. Turn the chicken four times. Forty minutes of cooking should be enough for a 2½ to 3 pound bird, but make sure that the flesh between thigh and drumstick is no longer pink. Thicken the sauce with egg yolk.

Variation: Prepare duckling the same way except remove all the fat from the pan as soon as it has browned and before adding the Vermouth.

Les pêches à la Capri (Peaches Capri). Boil a pint of red wine mixed with the juice of a lemon, 5 ounces of sugar and a pinch of powdered vanilla for 15 minutes. Place eight skinned peach halves in a salad bowl and sprinkle with the boiling wine. Let stand, then cool in a refrigerator. Serve with shaved ice.

Le sabayon (Zabayone). There are a thousand hot and cold versions of this famous dessert, but why not try one of the simplest . . . and the best. For a party of four, boil a pint of Marsala and 5 ounces of sugar for ten minutes. Beat six eggs into omelette condition. Pour in the liquid drop by drop, stirring constantly with a wire whisk. Pour the mixture into individual ramekins or a single mold and let it set over a double boiler. Serve ice-cold.

Germany

Coeur de veau en civet (Stewed Calf's Heart). Two hearts are adequate for four hearty eaters—no pun intended. Cleanse the hearts in several changes of cold running water so that they are completely free of blood. Remove inner and outer tendons and membranes; drain. Cut the hearts into quarter-inch slices and marinate for two days in wine—white or red—seasoned with onion, garlic, bay leaf, thyme, a carrot, parsley, a clove, two peppercorns and a sprig of pine. Melt 5 ounces of diced fat bacon in a stewpan and brown the meat in this. Dust with flour, salt, add the marinade and let the meat cook over a low flame for two hours. Just before serving, strain the sauce, thicken with a little heavy cream, bring to a boil and then cover the slices of calf heart.

Beef hearts can be prepared in the same way. In this case, however, I would advise you to marinate them for three days as the flesh is much less tender; and to count on an extra hour of cooking time.

Serve with unsweetened applesauce and a dish of buttered noodles.

Croûtes au vin (Breadcrusts in Wine, or German French Toast). Soak thick bread slices in red wine, to which sugar has been added, for ten minutes. Drain, then dip into egg whites beaten stiff, and brown gently in butter.

Serve piping hot, sprinkled with powered sugar and, if you wish, cinnamon.

Sauce chaude au vin rouge (Hot Red Wine Sauce) for rice puddings, etc. For each two glasses of red wine, add a pinch of cinnamon, one clove, a piece of lemon peel, four tablespoons of powdered sugar and a pinch of flour (mixed in a little water). Stir the sauce well over a moderate flame and serve after it has boiled five minutes.

Switzerland

Fondue. Typical and extremely popular in Switzerland, fondue is relatively unknown elsewhere except in neighboring Savoy and Jura. Nevertheless, it is winning new friends daily. It is simple to prepare, amusing to eat and relatively inexpensive. Traditionally, it is made in an earthenware crock with a handle, called a *caquelon*. Unfortunately this is on its way out, being replaced by a stronger enameled iron pot. The old earthenware crock was rubbed with garlic to keep it from cracking, and the same is done now to the enamel pot . . . just for tradition's sake.

Count on a quarter pound of cheese per person, choosing equal portions of Gruyère and Jura or Comté or Emmenthal. Cut the cheese into fine strips, place them in the pot and cover with dry white wine (just cover, no more). Put the pot over a medium flame and stir constantly with a wooden spatula. When the mixture is a lumpless liquid, add a small glass of kirsch and thicken with a pinch of flour. Pepper generously and carry the crock to the table where it should rest over an alcohol burner.

Beforehand, you will have provided each guest with a plate of slightly stale bread cut into inch-sized bites (about 30 pieces per person). Everyone then spears a morsel of bread with his fork and puts it into the common pot to soak up some cheese. Keep the flame of the alcohol burner rather high so that the fondue continues to cook as it is eaten. When the crock is empty, turn off the burner and scrape the crust of burnt cheese from the bottom.

True fondue adepts do not drink wine with this dish, but take an occasional small glass of kirsch. However, if you wish to serve wine, make sure it is comparable to the wine you used in cooking. If you are more than four at table, it is convenient to use two crocks.

La palée sauce neuchâteloise (Neuchâtel Lake Fish). A *palée* is a fish caught only in the lake of Neuchâtel; its cousin in Geneva is known as a *féra*.

Take one large fish or two medium-sized ones—small palées are rather soft-fleshed and are unsuitable for this dish although delicious fried or sautéd in butter. Scrape clean and place the fish in a court bouillon made of equal parts water and Neuchâtel wine which has cooked for an hour and been seasoned with onion, carrot, leek, parsley, lemon, salt and pepper (no garlic and no bay leaf).

Let the fish simmer for about 20 minutes. Check doneness with a knitting needle—it should penetrate the fish completely without effort. Meanwhile, make a white roux, moistened with a little of the court bouillon and with Neuchâtel. Check the seasoning of the sauce and, just before serving, thicken it away from the flame with an egg yolk and two tablespoons of cream. Bring once again to a boil and serve in a sauceboat. Serve the dish with steamed potatoes.

Using Vaud or Geneva wine, the same recipe is good for *féra à la vaudoise* or *à la genevoise*.

Variation: Large lake trout are also delicious prepared this way.

ZABAYONE, an Italian recipe (see page 301). *(Photo Chapman)*

La tarte au vin (Wine Tart). Line a pie tin within a half inch of the top with pastry dough. Dot with small chunks of butter and fill with a mixture consisting of a half cup of white wine, one whole egg, a tablespoon of powdered sugar and a pinch of salt, the whole well mixed with a whisk. Place in a hot oven for 25 to 30 minutes and serve hot, or at least warm.

Spain

La sangria. Spain's national drink varies in preparation from one region to another. I am giving you the formula I finally settled on which has the advantage of being thirst-quenching and of not going to your head . . . too fast.

Mix a quart of red wine with an equal amount of soda water, though the latter can be reduced if you like a stronger drink. Add a half cup of cognac (or gin if you prefer. This is less Spanish perhaps, but still good) and the juice of four oranges. Wash but do not peel an orange and a lemon and slice thinly. Dissolve four tablespoons of powdered sugar in a little water. Mix all the ingredients together. Add a pinch of cinnamon if you wish. Let the whole marinate for at least six hours, stirring from time to time. Sangria is marvelous with all summertime meals. It can be drunk either at room temperature or iced. If you use ice cubes, diminish the amount of soda water proportionately.

Potiron au vin (Pumpkin in Wine). Fry medium or large pumpkin slices in butter. Serve with a hot red wine sauce (for the sauce, see GERMANY).

Portugal

Strangely enough, the whole world seems to use Port and Malaga in cooking everything from soup to desserts. In their native Portugal, however, these wines are held in such respect that they are never used in any traditional dishes.

Le thon à la portugaise (Portuguese Tuna). Heat a little olive oil in a stewpan and brown a quarter pound of sliced onions. Make a roux, using a tablespoon of flour, then add a 2¼-pound piece of fresh tuna. Cover with white wine, add six quartered tomatoes, two cloves of garlic, thyme, bay leaf, salt and pepper. Let the whole simmer for two hours, occasionally turning the tuna.

Variation: Cut the fish into large pieces. Cook as above in a stewpan, adding peeled potatoes, cut into thick slices, to the other vegetables and spices. Let stand on the fire until the potatoes are completely cooked.

Add two large green peppers cut into thin slivers.

La crème au porto blanc (White Port Custard). In this custard, white Port is substituted for milk (red Port would result in a rather unappetising color). Boil a half bottle of Port with a little vanilla and a lemon twist. In a large bowl, beat stiff three whole eggs, five tablespoons of powdered sugar (¼ pound) and a teaspoon of flour. Pour the hot Port gradually over the sugar-egg mixture as you beat. Replace on the stove and continue beating until it thickens. The custard should not boil. Let it cool, and just before serving mix in a half pint of very heavy or whipped cream.

Serve garnished with macaroons dipped in Port.

Greece

Les filets de poisson à la grecque (Greek Fish Filets). Clean and filet one herring, mackerel or whiting per person. Place them in an oven-proof dish and cover with white wine to which has been added chopped parsley, onion, garlic, salt and pepper. Then add pitted green and ripe olives and place in a hot oven for 15 to 20 minutes. Serve with lemon slices.

Variation: Boil white wine combined with herbs and spices, salt and pepper for a quarter of a hour. Strain the marinade and add the juice of one lemon before pouring it boiling over the fish filets. Let the fish filets marinate for 48 hours and then serve them chilled accompanied by a side dish of chopped onion and parsley.

Champignons à la grecque (Greek Mushrooms). Use only fine small button mushrooms. Clean and wipe without washing if possible. For each pound of mushrooms, sauté a dozen small white onions and two cloves of garlic in a little olive oil. Add a half pint dry white wine, bay leaf, thyme, parsley, salt, pepper and the juice of one lemon. Let the mixture cook for ten minutes, then throw in the mushrooms for another 10 to 15 minutes. Serve cool, sprinkled with lemon juice.

Variation: Prepare artichoke hearts in the same manner.

Compote macédonienne (Macedonian Fruit Salad). Cook one pint of wine (white or red) with a half pound of sugar for ten minutes. Poach whole fresh figs in this for five minutes and grapes for two minutes. Remove this fruit from the syrup and then cook large cubes of melon for about five minutes in the same syrup until they are transparent. Add these to the other fruit. Reduce the syrup, then pour it over the fruit salad. Cool; serve ice-cold.

Variation: Add Corinthe raisins, which have been soaked in wine, to the compote while still hot.

United States

Le cocktail de crevettes (Shrimp Cocktail). Use frozen or canned baby shrimp. Wash them thoroughly in cold running water, drain and sprinkle with very dry white wine. Let them marinate for at least an hour, then drain a second time. Make a mayonnaise, replacing the vinegar with lemon juice and adding a few drops of white wine. Mix shrimp and mayonnaise together and serve chilled in cups or on a lettuce leaf.

Le jambon à la californienne (California Ham). Take four rather thick slices (about ¼ inch thick) of country ham and place them in a large frying pan. Cover with a half pint of dry white wine. Bring to a boil and let simmer for 20 minutes. Separately cook four peeled and pitted tomatoes in a little olive oil along with 5 ounces chopped mushrooms, very little salt and pepper, a little bay leaf and a pinch of basil. Thicken the sauce with egg yolk. Serve the ham rolled up on a platter and covered with the sauce.

Variation: Use cooked Virginia ham. Substitute red wine, Port or Malaga for the white wine. During cooking, add raisins which have previously been soaked in wine.

Les croûtes hawaïennes (Hawaiian Open Sandwich). Soak half-inch slices of slightly stale bread in dry white wine for ten minutes. Drain and sauté in butter. Spread one side of each slice with mustard, then place on top a slice of ham, a slice of drained canned pineapple and a slice of Swiss cheese. Place in a hot oven and serve when the cheese is melted and browned. Season to taste with paprika, nutmeg or simply pepper.

Variation: Replace the pineapple slice with a slice of orange.

ACKNOWLEDGMENTS

The authors of THE WINE BOOK owe a great deal of the information used to those listed below, and would like here to express their sincere thanks.

FRANCE
Arnaud, Claude, Lunel, Hérault
Bolter, William, Bordeaux
Boulat (M.), Reuilly, Indre
Breval, Stanislas, Autremencourt, Aisne
Chandon-Moët, Count Raoul, Moët & Chandon Champagne, Epernay, Marne
Chevillard (M.), grand chancellor, Chevaliers du Tastevin, Clos Vougeot, Côte d'Or
Brugirard, André, director, Wine Research Centre of the Agricultural Chamber of Pyrénées-Orientales
Creuilly (Mlle.), International Wine Office
Croislay, Georges, La Motte, Var
Flament (M.), warning station, Beaune, Côte d'Or
Fleury (Count and Countess), Etablissements de Neuville, Saint-Hilaire-Saint-Florent, Maine-et-Loire
Gallay, Violette, La Roche-aux-Moines, Maine-et-Loire
Hogmar, H.N., Etablissements Cordier, Bordeaux
Jaboulet Verchère (M.), Pommard, Côte d'Or
Lanson, N., Lanson Champagne, Reims, Marne
Latour, Louis, Beaune, Côte d'Or
Léglise, Professor Max, director, Wine Research Centre, Beaune, Côte d'Or
Lelorrain (M.), Comité interprofessionnel des vins de Bordeaux (C.I.V.B.), Bordeaux
Malestroit (Count de), Vallet, Loire-Atlantique
Miramon (M.), Union for the Defence of Jurançon Wines
Mommessin, Jean, Mâcon, Saône-et-Loire
Orizet, Jean, Geneva, Switzerland
Pasquier-Desvignes, Claude and Marc, Saint-Lager, Rhone
Pellenc-Turcat, F., Paris
Peynaud, Professor E., chief of research, Agronomic and Wine Research Centre, Bordeaux
Rasque de Laval (Baron de), founder, Côtes de Provence Union and Ordre des Chevaliers de la Méduse
Rochecouste (Mme. de), Savennières, Maine-et-Loire
Rohan-Chabot (Count de), president, Union for the Defence of Côtes de Provence, Var
Sanders, J., Château Haut-Bailly, Léognan, Gironde
Schaeffer, J.A., Strasbourg, Bas-Rhin
Seltz, A., Seltz & Fils, Mittelbergheim, Bas-Rhin

OTHER COUNTRIES
Antinori (Marquis), Florence, Italy
Ayuso Murillo, Antonio, president, Regulating Council for the denomination "Mancha", Spain
Bartlett, Jonathan, New York, United States
Berdejo Siloniz, Salvador Ruiz, president, Regulating Council for the denomination "Jerez-Xeres-Sherry", Spain
Bibi (M.), secretary general, Tunisian Wine Cooperatives, Tunis
Bourquin, Constant, president, Swiss Wine Academy, Sion, Valais, Switzerland
Broussilovski, S., director, Champagne Factory, Moscow, U.S.S.R.
Callegari, Danilo, Companhia Vinicola Rio Grandense, Brazil
Campuzano (D.), Puerto Real, Spain
Chicote, Pedro, Madrid
Dornier, R.P., Domaine de Thibar, Tunisia
Dreher Neto, Carlos, Brazil
Dalmasso, Professor G., Italian Vine and Wine Academy, Italy
Dubois (MM.), Epesses, Vaud, Switzerland
Dülger, Mehemet, Geneva, Switzerland
Fould, G.S., Australian Wine Centre, Australia

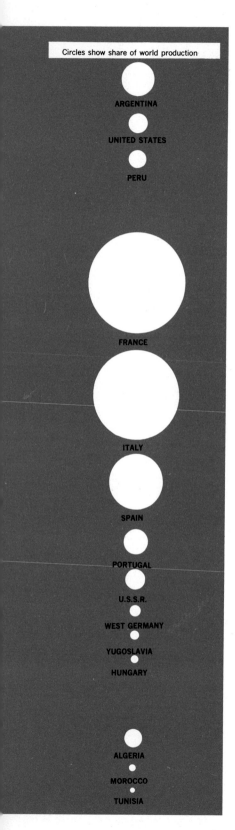

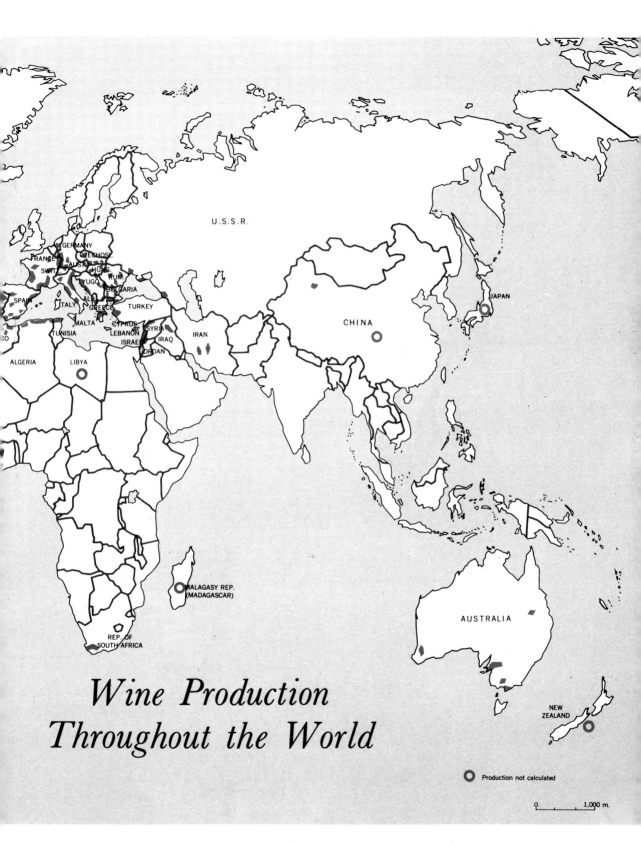

Wine Production Throughout the World

Ghinst, R. Van der, Ksara Cellars, Lebanon
Gouvea de Freitas, Gabriel P., Brazil
Groues, Henri, Mexico, D.F., Mexico
Imhof, Louis, former president, Swiss Wine Academy, Sion, Valais, Switzerland
Jardi Porres, Alberto, general manager, National Winemakers Association, Mexico
Jelaska, Marcel, Vine Culture Institute, Split, Yugoslavia
Kok, Einar, São Paulo, Brazil
Laden, Norman, San Francisco, California, United States
Matuschka-Greiffenclau (Count), president, Union of German Winegrowers, West Germany
Michaux (M.), director, Federation of Valais Wine Producers, Sion, Valais, Switzerland
Noireau, J., Algiers, Algeria
Ortiz Garcia, His Ex. M. Nazario, Mexico
Pujades de Fria, Luis, president, Regulating Council of the denomination "Tarragona", Spain
Ratti, Renato, Cinsano, Italy
Rensburg, L.J.J. van, and Botha, P., Vintners' Cooperative Association of South Africa
Rosenthal, P.S., Grandes Caves Wine Cooperative Company, Rishon-le-Zion, Israel
Ruiz-Campuzano, Balbina, Madrid, Spain
Shile, W., Pleasant Valley Wine Company, United States
Schubert, Baron Andreas von, Trier, West Germany
Schloss Vollrads (H.), Winkel-im-Rheingau, West Germany
Sté Vinicole Los Ranchos, Uruguay
Tekel Genel Müdürlügü, T.C., Istanbul, Turkey
Traldi, Alberto, Etablissements Traldi, São Paulo, Brazil
Undurraga Fernandez, Pedro, Viña Undurraga, Brazil
Vidal Barraquer, José, president, Regulating Council for the denominations "Alella" and "Panades," Spain
Vidal Barraquer, Ramon, president, Regulating Council for the denomination "Priorato", Spain
Wagner, Philip, Boordy Vineyard, Riderwood, Md., United States
Wuilloud, Charles, and Mme. Wuilloud, Sion, Valais, Switzerland

PROPERTIES
Glasses graciously lent by Luce, Paris
Wines from Auge, Paris

DOCUMENTATION
Offidoc

INDEX

Names of persons are in small capital letters. Asterisks indicate illustrations.

A

Abrau-Dyurso, 231*
Aconcagua, 267–68
Adelaide, 276
ADLUM, JOHN, 259–60
ADRETS, BARON DES, 133
Adriatic, 244–46
Aegean Islands, 29–30
Aeolian Islands, 165–66
AGABALIANTZ, GREGORI, 232–33
AGAMEMNON, KING, 22
Ahr Valley, 171
Ak-Tepe, 18
Alameda County, 252, 257–58
Alava, 197
Alba, 33, 155
ALCIBIADES, 24
ALDRICH, HENRY, 103
Aleksandrovac, 244
ALEXANDER THE GREAT, 30, 228
ALEXANDER OF TALLES, 30
Alexandria, 207*
ALFONSO, JORGE, 204
Algeria, 208–11, 211*
Algiers, 209
AL-HAKIM, 218
Alicante, 196
Almadén, 252–55
Aloxe-Corton, 103
Alsace, 51*, 59*, 139–43, 139*, 142*, 143*, 299–300
Alsheim, 179
Amboise, 126
Ambonnay, 114, 116
American Revolution, 77
AMERINE, MAYNARD A., 62
Amiata, Mount, 153
Ammerschwihr, 142
Amontillado, 192
Ampuis, 131
Anatolia, 223
Andalusia, 191
Andlau, 143
Angers, 129
Anjou, 57, 71, 128–29
Ankara, 223
Anniviers Valley, 187
Antibes, 34, 165
ANTINORI, GIOVANNI DI PIERO, 151
ANTINORI, NICOLO, 152
ANTINORI, PIERO, 152
Antioch, 223*
Antioch Chalice, 39*
ANTIPATER OF TYRE, 24
Antipolis, 34
Apulia, 163–64
Aquitaine, *see* Bordeaux
AQUITAINE, ELEANOR OF, 70
Ararat, Mount, 18
Arbia river, 153
Arbois, 146
Argentina, 265–67, 265*
ARIADNE, 24
ARISTOTLE, 33–34
Armenia, 18, 228
Assmannshausen, 177
Asti, 155, 255
Astrakhan, 229
Atacama, 268

ATATÜRK, KEMAL, 223
Athens, 23–24, 30, 216
Atlantis, 216–17
AUGUSTUS, 167
Auslese, 170–71
AUSONIUS, 15, 87
Australia, 274–76, 274*, 275*
Austria, 189–90, 189*, 190*
Auxerre, 71
Avdat, 221
Avenay, 114
Avignon, 131
Avize, 114
Ay, 112, 114, 116
Azay-le-Rideau, 126
Azé, 98

B

Bacchante, 245*
Bacchanal, 23*, 26*
BACCHUS, 6*, 31*, 214*, 215*
Bacharach, 168, 177
Bactria, 228
Badacsonyi Hills, 239
Bad Durkheim, 179
BADE, GUILLAUME DE, 58*
Baden, 180, 190
Balaton, Lake, 239
Balatonfüred-Csopak, 239
Baltimore, 262
BALZAC, HONORÉ DE, 126, 134
BARBAROSSA, FREDERICK, 133
Baricska, Csarda, 239
Barr, 143
Barrossa Valley, 275
Barsac, 92
BARTHOLOMEW, FRANK, 255
Barton et Guestier, 79
Bas-Rhin, 140, 143
Bastide Blanche, 135
Bâtard-Montrachet, 101
Battle of Castillon, 74
Battle of Nuits, 109
Bavaria, 172
Beaujolais, 50–51, 61, 63*, 64, 93–99, 95*, 97*, 99*
Beaulieu Vineyard, 256
Beaune, 102–03, 112
Beblenheim, 143
Beerenauslese, 171, 176
Belgium, 206
Belgrade, 244
BELMONTE, JUAN, 194*
BENOIST, LOUIS, 252
BERCHIER, 111–12
BERENGARIA, QUEEN, 217
Bergerac, 145
Bergère-les-Vertus, 114
Beringer Brothers, 256
Bernkastel, 148–49*, 168, 180
Berry Brothers, 76–77
Bessarabia, 241
Bethlehem, 218
Bèze, Abbaye de, 108
Bible, 16–18, 17*, 20–21, 21*, 36, 218
Bienne Lake, 186
Bingen, 177–79
Bisagno Valley, 156

Bisevo, 245
BISMARCK, 272
Bissey, 100
BISSON, COLONEL, 105
Black Forest, 180
Black Sea, 227
BLAXLAND, GREGORY, 275
Bodenheim, 179
Bohemia, 241
Bommes, 92
BONAPARTE, NAPOLEON, 105, 107*, 108–109, 147, 176, 194*, 205, 271, 273
Bône, 210
BONNARD, 113*
Bonnes-Mares, 108
Book of Hours of Guillaume de Bade, 58*
Boordy Vineyard, 262
Bordeaux, 61, 64, 67–92, 72–73*, 81*, 83*, 85*, 86*, 293–96
Bordelais, 67
Bosnia, 242*
Bottling, 64–65
Botrytis cinerea, 88–91, 90*, 125, 171, 235, 258
Bougie, 210
Bourgueil, 127–28
BOUSCHET, LOUIS, 137
Bouzy, 114, 116
BOZILLE, GASTON, 137
Brac, 245
Brazil, 268–70, 269*
Brda, 244
BRENTEL, FRÉDÉRIC, 58*
Bresse, 96
Bristol, 70
Brittany, 65*
BROSSE, CLAUDE, 98–99
Brouilly, 94
Bruderberg, 182
Bruley, 144
BRUSSILOVSKY, SERGEI, 233
BRUYN, GODFREY DE, 284
Bucovina, 241
BUDDHA, 19
Bué, 125
Buena Vista, 250*, 255
BUGEAUD, THOMAS, 209
Bulgaria, 240–41
BULL, EPHRAIM, 260
"Bull's Blood," 239
BURCHARD, 218
BUREAU, *Maître*, 90–91*
Burgenland, 190
Burgos, 197
Burgundy, 43, 51, 57, 61, 66–67*, 93–110, 95*, 97*, 98–99*, 100*, 103*, 277*, 295*, 296–97
BUSBY, JAMES, 275
Buxy, 99

C

Cádiz, 191
Cadre Noir, 128
CAESAR, JULIUS, 33, 160, 167
Cahors, 145
Caiolo, 151
Cairo, 222
Calabria, 164

California, 49, 61, 199, 247*, 249–59, 250*, 253*, 254*, 257*, 258*, 264
Canaan, 20, 21*, 218
Canada, 243, 259, 262
Canary Islands, 198
Canelones, 270
Canossa, 160
Canthare, 26*
Cap Corse, 147
Cape Province, 272
Capri, 162–63
CARAVAGGIO, MICHELANGELO, 6*
Carbonic maceration, 58
Carmel Winegrower's Cooperative, 219
Carpathians, 241
Casablanca, 213
Casa Vinicola Barone Ricasoli, 151, 153
Castellina, 151
Castello di Brolio, 153
Castelnuovo Berardenga, 151
CASTILLO, DON PEDRO DE, 266
Catina, 165
CATO THE ELDER, 32
CATULLUS, 33
Caxias do Sul, 268–69
Cecubis, 33
Cerignola, 164
Chablais, 185
Chablis, 61, 93, 110
Chaintré, 98
Chaldea, 20
Chalon, 99
Chambertin, 107*, 108
Chambertin Clos-de-Bèze, 108
Chambolle-Musigny, 107–08
Chambre des Députés, 129
Champagne, 51*, 61, 111–21, 111*, 113*, 114*, 117*, 118–19*, 120*, 231–34, 231*, 297
Champagne Vineyard, 112–14
Champanskoye, 231*, 232–34
Champigny, 128
Chaouia Plains, 213
CHAPTAL, JEAN, 59
Chaptalization, 59
Chardonnay, 95, 98, 100
CHARLEMAGNE, 38–39, 159, 167–68, 176
CHARLES V, 165
Charles Krug Winery, 256
CHARLES VII, 125–26
CHARLES X, 109
Chartrons, Quai des, 79
Chassagne-Montrachet, 101
Chassagne-Montrachet Grandes-Ruchottes, 101
Château Aloxe-Corton, 100*
Château Ausone, 86*, 87
Château Carbonnieux, 87
Château-Châlon, 146, 187
Château Cheval-Blanc, 87
Château du Clos-Vougeot, 106*
Château of Corton-Grancy, 100*, 104
Château Grancy, 103–04
Château Grillet, 131
Château Haut-Brion, 85*, 86
Château Latour, 84
Château-Lafite, 81*, 84
Château Lafite-Rothschild, 81–83
Château Masar, 221
Château Mouton-Rothschild, 85
Châteauneuf-du-Pape, 58, 133–34
Château Sainte-Roseline, 135*
Château-Thierry, 112
Château de Versailles, 98–99
Château Yquem, 89–92, 90*, 91*
Chavignol, 125
CHAUCER, GEOFFREY, 73, 166
CHAUVERT, JULES, 107, 283

Chénas, 94
Chenon, 126–27
Chenonceaux, 126
Chevalier-Montrachet, 101
Chevaliers du Tastevin, 107
Chianti, 151–54
Chile, 266*, 267–68
China, 19
Chincha, 270
Chios, 29–30
Chiroubles, 94
Christian Brothers, 257
Christianity, 36–39
Chrysatikos, 30
CICERO, 162
Cienega, 253
Cinecitta, 162
Cinque Terre, 156–58, 163*
Clessé, 98
CLIQUOT, MADAME, 118
Clochemerle, 95*
Clos-des-Lambrays, 108
Clos-Saint-Denis, 108
Clos-Saint-Jacques, 94
Clos-de-Tart, 108
Clos-Vougeot, 61, 105, 108
Coblenz, 177, 180
COHENS, 220
Colares River, 204
COLBERT, JEAN, 74
Colchagua Province, 267
Colchis, 227
COLETTE, 144
Cologne, 177
Colmar, 142–43
COLUMELLA, 33, 161
Comblanchien, 104
Comité interprofessionel des vins de Bordeaux (C.I.V.B.), 77–79
Comité interprofessionel des vins de Champagne, 121
CONCANNON, JAMES, 258
Concannon Vineyard, 258
Concord, 260
Condrieu, 131–32
Confrérie des Chevaliers de Méduse, 136
Confrérie des Sacavins, 129
Confrérie Saint-Etienne d'Alsace, 142
Confrérie des Tastevins, 105
CONFUCIUS, 19, 290
Constance, Lake, 180
Constantia Valley, 272
CONTI, PRINCE DE, 105
Cooperative Wine Growers Association of South Africa, Ltd., 272
Coquimbo, 268
Cordier, 79
Córdoba, 196
Corfu, 216
Corks, 114*, 119–21
Corkscrews, 290–92, 292*
Cornas, 133
Corsica, 147
CORTES, HERNANDO, 271
Corton, 103
Côteaux de la Loire, 129
Côte d'Azur, 136
Côte de Beaune, 93, 100–09
Côte des Blancs, 114
Côte de Brouilly, 94
Côte Chalonnaise, 93, 99
Côte de Nuits, 93, 104–09
Côte d'Or, 93, 100–09
Côtes de Provence, 134–38, 135*, 138*
Côtes du Rhône, 131
Cotnari Winery, 241
Coulée-de-Serrant, 129
Coupage, 79
Cramant, 114

Cres, 245
Crésancy, 125
Cresta Blanca Vineyards, 257–58
Crete, 216
Crimea, 229*, 231
Croatia, 242–43, 246
Croix Valmer, 135
Crozes Hermitage, 132*, 133
Cruse, 79
Cucamonga, 259
Curicó Province, 267
Cuyo, 266
Cyprus, 217
CYRUS, 34
Czechoslovakia, 240*, 241–42

D

DAGOBERT, KING, 142
Dalmatia, 244–46
DANIEL, JOHN, 264
DANTE, 166
Danube river, 243
Dao Valley, 204
Daphni, 216
DAVID, KING (Scotland), 70
Deidesheim, 179
Delos, 34
Delphi, 22
DEMARGUE, PIERRE, 34
Denicé, 94
Derrière-la-Grange, 108
Dézaley, 184–85
Die, 133
Dienheim, 179
Dijon, 109
DILLON, DOUGLAS, 82
DIONYSUS, 22–24, 30, 215
Disgorging, 119
DJEMSHEED, SHAH, 224
Djidjelli, 210
Doktorberg, 180
Dôle, 185
DOMITIAN, 35–36
Dordogne, 67, 145
Dosing, 119
Doukkala Plains, 213
Douro River valley, 199–203, 203*
DRAKE, SIR FRANCIS, 195
Drava river, 243
Drawing, 84
Drôme, 59*, 133
Drosselgasse, 176*, 179*
DU GUESCLIN, BERTRAND, 84
DUMAS, ALEXANDER, 101, 108
Dürnstein, 190
Dutch East India Co., 271

E

Eastern Europe, 227–46
Eberbach, 168
Echézeaux, 105
Edelbeerenauslese, 171
EDWARD II, 72
EDWARD III, 70
Eger, 239–40
Eguisheim, 142
Egypt, 18*, 19, 20*, 21*, 22, 208*, 220*, 222
Eiswein, 171–72
Eitelsbacher-Jarthäuserhofberg, 181
E. & J. Gallo Winery, 264
El Alamein, 222
Elbe Valley, 240*, 242
El Cerro Hill, 270
England, 276*
Entre-Deux-Mers, 88
Eperjes-Tokaj Mountains, 235

Epernay, 115–16, 121
Epesses, 184–85
Erech, 16
ERICSON, LEIF, 249
ERIC THE RED, 249
Eschenauer, 79
ESQUIER, LOUIS, 269
Estoril, 204
Estramadura, 204
Etablissement Nicolas, 138
Etna, Mount, 165

F

FALSTAFF, 191, 205
Fargues, 92
Farsistan, 19
Fattorie Antinori, 151–52
Federweisser, 172
FERDINAND, EMPEROR, 249
Fermentation, 53–54
Festivals, *see* Wine festivals
Fez, 213
Fiaschi, 152–53*
Ficklin Vineyards, 259
Finger Lakes, 260–61
Fino, 191–92
FITZGERALD, EDWARD, 224
Flagey-Echézeaux, 105
Fleurie, 94
Flor, 192, 194, 272
Florence, 151, 153–54
Forst, 179
FOURNIER, CHARLES, 262
France, 66–149
Franche-Comté, 146, 300
Franconia (Franken), 183
Frankenthal, 174
Franschhoek, 271
Frascati, 162
Fresno County, 259
Frontignan, 138
FUGGER, JOHANNES DE, 161–62
Fuissé, 98

G

Gaillac, 145–46
GALLO, ERNEST, 264
Gallo Nero, 152
Garda, Lake, 159
Garonne river, 67
Gascony, 67, 72
Gassin, 135
Gaul, 33–36, 35*
GAUTIER, THÉOPHILE, 192
Geneva, 184–86
Georgetown, 259–60
Georgia (Soviet), 227, 230*, 231–32
German Riviera, 174
Germany, 71, 148–49*, 167–83, 167*, 170*, 171*, 172*, 173*, 175*, 181*, 182*, 183*, 301
Gevrey, 109
Gewürztraminer, 143
GIANACLIS, NESTOR, 222
Gibraltar, 191
Gironde, 67, 80, 82, 84, 86*, 87
Givry, 99
Glacier wines, 187–88
GLADSTONE, 77, 272
Gloucestershire, 38
GOBLET, DOM, 107
God of Wine, 22–23
Gods, 6*, 18, 22–24, 30, 31*, 161, 215*, 229
GOERING, HERMANN, 84
GOLDBERG, 178
Gold Seal Vineyards, 262

Gönc, 61
Gonfaron, 135
Gonzalès and Byass, 193*, 194*
Gordo, 192
Goutte-d'Or, 101
Graach, 180
Graisse, 117
Grands-Echézeaux, 105
Grandes-Ruchottes, 101
Grape, anatomy of, 52–53
Grape growing, 40–41*, 43–48, 46*, 47*, 92*, 123, 129–30, 152, 215, 247*, 261*, *see also* Harvesting; Wine making
Grasse, 136
GRATIAN, 87
Graves, 80–81, 86–87
Great Hungarian Plain, 236
Greece, 215–17, 217*, 305
Greece, classical, 22–24, 23*, 26–30, 26*, 27*, 29*, 33–34, 214*, 215*
Greenland, 249
Greenland Saga, 249
Greve, 151
Grimentz, 187
Groot Constantia, 271–72
Guebwiller, 142
Guerneville, 255
Gulf of Naples, 162–63
GULNARE THE FAIR, 224
Gumpoldskirchen, 190

H

Haardt Hills, 179
HABISHAM, RAINWATER, 205
HAFIZ, 19
Hallenheim, 177
Hallgarten, 177
Han Dynasty, 19
HANNIBAL, 160
HARASZTHY, AGOSTON, 249–51, 250*, 255, 275
Haro, 197
Harvesting, 20*, 37*, 48–51, 48–49*, 51*, 52*, 56*, 58*, 59*, 88–89, 95–96, 95*, 110*, 114–15, 117*, 128*, 136*, 141*, 142*, 160*, 163*, 183*, 200–01, 213*, 229*, 253, 263*, 267, 273*, *see also* Grape growing; Wine making
Haut-Rhin, 140, 143
Hautvilliers, Abbey of, 111–12
Hegyaljai, 235
Heidelberg, 169*
Heiligenstein, 143
Hellenic growths, 29–30
HEMINGWAY, ERNEST, 105
HENRI IV, 144, 146
HENRI III, 74
HENRY VIII, 195
HENRY IV, 99, 112, 127
HENRY II, 70–72
HENRY III, 72, 87
HENRY THE NAVIGATOR, PRINCE, 198
HERA, 215
Hermitage, 132*
Herrenberg, 178, 182
Herzegovina, 245
HESIOD, 215
Hesse, 179
HESTIA, 23
Heurige, 189*, 190
Hittites, 21–22
Hochheimer, 168, 173*, 177
Hohwald, 143
HOMER, 22, 28–29, 227
Homeric wines, 28–29
HORACE, 33, 162

HORTON, DONALD, 16
Hospice of Beaune, 103*
Hôtel Dieu, 102*
HUGH IV (Cyprus), 70
Hunawihr, 139*, 143
Hundred Years' War, 74, 84
Hungary, 225*, 226*, 234–40, 235*, 236*, 237*, 238*
Hunter River valley, 275
Hvar, 245, 246

I

Ica, 270
Icaria, 23
ICARUS, 31*
Igé, 98
Ile de France, 112–14, 271*
Ill River, 139
Imotski, 246
Inaouene River, 213
India, 19
Ingelheim, 167
Inglenook Vineyard Co., 256, 264
Ionia, 34
Iran, 18–19, 224
Isaiah, 218
Isca, 206
Ischia, 163
Ismarus, 28–29
Ispahan, 224
Israel, 218–21, 218*, 219*
Istria, 244–46
Italian Swiss Colony, 255–56
Italy, 40–41*, 151–66, 157*, 160*, 163*, 300–01, 303*
Ivriz, 18

J

Jaffa, 219
Japan, 259
JAPHETH, 205
JAY, DOUGLAS, 255
JEFFERSON, THOMAS, 259
JEREMIAH, 20–21
Jerez de la Frontera, 191, 193*, 194*, 195
Jerusalem, 220
Jeruzalem (Yugoslavia), 243
JESUS, 36
Jews, 20–21
Jordan, 222
JOSHUA, 21*
JOAN OF ARC, 126
Johannesberg, 168
JOHN, KING (England), 70
JOHN, KING (France), 70
Judea, 219*
Juliénas, 94
Jully-les-Buxy, 99
Jundiai, 269
Jungferwein, 172
JUPITER, 161
Jura, 146
Jurades de Saint-Emilion, 280
Jurançon, 144–46

K

Kabirion, 26*
Karlovo, 241
KELLERMANN, 176
Kern County, 259
Kfar Cana, 218
KHAYYAM, OMAR, 19, 224
King County, 259
KISFALUDY, SANDOR, 239
Kokinelli, 216

Korbel Winery, 255
Korcula, 243*, 245
Kottabos, 26-28, 27*
Kramolin, 240-41
Krasnodar Alimentary Institute, 232
KRAU, EDOUARD, 39
Kresmann, 79
KRESSMAN, EDOUARD, 284
Kreuz, 178
Krk, 245
Kröttenbrunner, 178
Kues, 168
Kurfursten Hoff, 169*
Kvarner, 246
Kyathos, 29*

L

La Chapelle de Guinchay, 96
LACKLAND, JOHN, 71
La Confrérie de la Chantepleure, 127
Ladoix-Serrigny, 104
LAFAYETTE, MARQUIS DE, 77
Lalande, 87-88
Langeais, 126
Langhe Hills, 155
Langon, 67
Languedoc, 136-38
La Romanée, 101
La Salle, Mont, 257
La Salle Military Academy, 257
La Spezia, 156
Lastovo, 245
La Tache, 105
Latin America, 265-71
Latium, 34
LATOUR, GEORGES, 256
LATOUR, LOUIS, 103
Laubenheim, 179
Lausanne, 184-85
Laval, 38*
Lebanon, 221-22
LEFRANC, CHARLES, 252
Leman, Lake, 187*
Le Mesnil-sur-Oger, 114
Lemnos, 30, 216
Lepe, 73
LEPIDUS, EMILIUS, 156
Les Amoureuses, 108
Les Arcs, 135
Lesbos, 30
Les Charmes, 101, 108
Les Genevrières, 101
Les Pucelles, 101
L'Etoile, 146
Levant (Spain), 196
Levanto, 156
Leynes, 96
Libourne, 87-88
LICHINE, ALEXIS, 79
Liebfraukirche, 178
Liguria, 156-58, 157*
Lima, 270
Lipari Islands, 165-66
Lisbon, 204
Liverdun, 144
Livermore Valley, 252, 257-58
Logroño, 197
Loire Valley, 51, 122-30, 124-25*, 127, 277*, 297
LOISIER, DON JEAN, 107
Long Island, 260
Lontué Valley, 267
Lorgues, 135
Lorraine, 144
Los Angeles, 249
Los Flores, 250
LOUIS XI, 102
LOUIS XII, 74
LOUIS XIV, 82, 98-99, 237-38
LOUIS XV, 82, 121
LOUIS XVIII, 109
LUCIFER, 162
LUCULLUS, 162
Lugny, 98
Lunel, 138
LUR-SALUCES, MARQUIS DE, 89-90
Luxembourg, 180, 206
Lyon, 94, 131
Lyonnais, 277*

M

MACARTHUR, JOHN, 275
Macedonia, 242, 244
Mâcon, 96, 98
Mâconnais, 93, 96-99
Madeira, 205
Madera, 259
Madiran, 145
Madrid, 197
Madrid Pact, 221
MAGELLAN, 240
MAIANO, GIULIANO DA, 152
Mailly, 114
Main River, 177, 183
Maipo Valley, 267
Maison Mâconnaise des Vins, 98
Maiwein, 172
Málaga, 196
Malmesburg, 272
MANDELLI, LOUIS, 269
Manhattan College, 257
Massilla (Marseilles), 34
Marches, 160
Marengo, 105
MARESCALCHI, ARTURO, 155
MARGUERITE OF BURGUNDY, 108
MARIA THERESA, QUEEN, 238
Maribor, 243
Marienthaler, 190
MARINO, 160
Marmoutiers Abbey, 126
Marne river, 111*
Marrakech, 213
Marsala, 165
Marsillargues, 138
MARTIAL, 33
MARTINI, LOUIS, 252, 256
Marseilles, 136
Maryland, 262
Massachusetts, 260
Massandra, 229*
Massico, Mount, 33
Massif Central, 136
MATUSCHKA-GREIFFENCLAU, COUNT, 174*
Maximin-Grünhause, 181-82
Mayacamas Hills, 250-56
Médoc, 64, 80-85
Mehun-sur-Yèvere, 126
Meknès, 213
Mělnik, 242
Memphis, 222
Mendocino County, 252
Mendoza, 265*, 266-67
Mercurey, 99
Mesopotamia, 18-20
METTERNICH, PRINCE, 176
Meursault, 101
Mexico, 270*, 271
Michigan, 259
Middle Haardt, 179-80
Middle Rhine, 177-80
Midi, 136-38, 198-99
Milan, 158, 160
Milly, 110
Minho, 202-05

Mission Dolores, 250
Mission San Diego, 249
Missouri, 259
MITHRA, 22
Mitidja, 210, 211*
Mittelwihr, 143
Modesto, 264
MOËT, CLAUDE, 121
Moët et Chandon, 116
Moldavia, 229
Moldavian Plain, 241
Molsheim, 143
MOMMESSIN, JEAN, 108
Monaco, 34
Monbazillac, 145
Moncontour, 126
Monemvasia, 217
Monferrato, 155
Monoïkos, 34
Montagny, 99
MONTAIGNE, MICHEL EYQUEM DE, 92
Montefiascone, 161-62
MONTEZUMA, 271
Montilla, 192
Montlouis, 126
Montmelas, 94
Montpellier, 137-38
Montrachet, 101
Montrésor, 126
Montrichard, 126
Mór, 239
Moravia, 241
Morey-Saint-Denis, 108
Morgon, 94
Morocco, 212-13*
Moscow, 227, 232
Moscow Champagne Factory, 232-34
Moselle, 39, 50, 61, 144, 148-49*, 167*, 168, 180-83, 182*
MOSES, 20, 218
Moulin-à-Vent, 94
Mulhouse, 140
MUNCHHAUSEN, BARON, 142
Mura River, 243
Murcia, 196
Muscadet, 129-30
Muscat, 205
Musigny, 108

N

Nackenheim, 179
Nahe River, 177, 182
Nantes, 129
Napa County, 252, 256
Napa Valley, 256-58
Naples, 31*, 162
Napoleon, *see* Bonaparte, Napoleon
Narbonne, 35
National Association of Viticulturists and Viniculturists, 271
National Institute of Appellations of Origin (I.N.A.O.) 49, 81
National Institute of Port Wines, 201
Naurazga-Tepe, 18
Navarra, 197
Naxos, 30, 215-16
Nérac, 87-88
NERO, 31
Neustadt, 179
New England, 249
New South Wales, 274-76
New York State, 259-62, 260*, 261*, 263*
Niagara Peninsula, 262
Nicaragua, 250
Nice, 34
NICHOLAS II, 195

Nicolas, 138
Nierstein, 178
Nikai, 34
Nile River, 222
Nîmes, 138
Nimrod, 13*
NINEVEH, 18
NOAH, 16–18, 17*, 205
NOISOT, 109
Nordeste, 269
North Africa, 209–14
Nuits-Saint-Georges, 104–05
Nysa, 215

O

Obernai, 143
ODYSSEUS, 22, 28–30
Oestrich, 177
O'Higgins Province, 267
Ohio, 259
OMBIAUX, MAURICE DES, 101
Omodhos, 217
Oporto, 202–03*
Oppenheim, 170*, 178
Oran, 209–10
ORIZET, LOUIS, 280, 283–84
Orla, 219–20
Orléans, 71, 125
ORPHEUS, 22
Orvieto, 161
OSIRIS, 18
Osoye, 73
Ott, Domaines, 136
Oujda, 213

P

Paarl, 272
Paicines, 252–53
Palatinate, 168, 179–80
Palestine, 20–22, 218
Palermo, 33
PARACELSUS, PHILIPPUS, 234
Parramatta, 274
Paros, 216
Parthenon, 23
PASTEUR, LOUIS, 53, 55*, 63–64, 146
Patmos, 216
Patrimonio, 147
Pau, 145
Pauillac, 82, 84
Paul Masson Winery, 247*, 254*, 257*, 258–59
PAUSANIAS, 24
Pedro Domecq, 194*
PEDRO THE SECOND, 268
Pelcera Valley, 156
Peljesac Peninsula, 245
Peloponnesus, 216
PEPI II, 222
PÉRIGNON, DOM, 111–12, 116, 121
Périgord, 67
Pernand-Vergelesses, 104
Persepolis, 224
Persia, 19, 25*, 224
Persian Gulf, 16
Perth, 275
Peru, 270–71
PETER THE GREAT, 238
Petite Montagne de Reims, 116
PEYNAUD, 285–87
Pfalz, 175*
PHILIPPE, LOUIS, 109, 272
PHILLIP, ARTHUR, 274
Phillippeville, 210
Phocaea, 34
Phylloxera, 44, 137*, 197, 209, 212, 251

Physicians' Wine Appreciation Society, 280
PICARD, HENRY, 70
PICHARD, SIEUR, 82
Piedmont, 154–56
PIERRE, BROTHER, 111–12
Pierrefeu, 135
Piesport, 180
PLANCHON, JULES, 137
PLATO, 23–24
Pleasant Valley, 263*
Pleasant Valley Winery, 262
PLINY THE ELDER, 33, 161
PLUCHE, ABBOT, 112
Poinchy, 110
Poitou, 277*
Poligny du Jura, 146
POLO, MARCO, 224
POLYBIUS, 32
Pomerol, 87–88
Pommard, 102
POMPADOUR, MADAME DE, 82, 105, 121
Pompeii, 162
POPE BENEDICT XIV, 238
POPE CLEMENT V, 86–87
POPE JOHN XXII, 133
POPE MARTIN IV, 166
Po River, 155
PORRES, ALBERTO JARDÉ, 271
Port, 74, 199–203, 200*, 202–03*
Portugal, 61, 73–74, 199–205, 200*, 202–03*, 304
Pouilly-Fuissé, 96, 98
Pouilly-Loché, 98
Pouilly-sur-Loire, 123
Pouilly-Solutré, 98
Pouilly-Vinzelles, 98
POUSSIN, NICOLAS, 218*
Prague, 241
Prater, 190
Preignac, 92
Première Côtes de Bordeaux, 88
Priorato, 197
PROBUS, 36
Prohibition, 251
Provence, 134–38, 135*, 138*, 298–99
Pruning, 43–46*, 92*, 261*
Puerto de Santa Maria, 192
Puligny-Montrachet, 101
PUSHKIN, 232
Pyrenees, 136

Q

Queensland, 275
Quincy, 126

R

Rabat, 213
RABELAIS, 127, 146
Racodium cellare, 82–84, 104
Radgona, 243
Ramatuelle, 135
RAMSES III, 19, 20*
RAY, MARTIN, 259
Recoltant manipulant, 121
Redda, 151
Red wine production methods, 57–60
Reggio, 164
Reims, 115–16, 121
Reims, Mount, 114
Religion and wine, 16, 22–23, 36–39, 218–20
Remuage, 118–19
Retsina, 216
Reuilly, 126
Rharb, 213
Rheingau, 168, 174–77

Rheinhessen, 178
Rheinpfalz, 179–80
Rhenish Hesse, 178
Rhine Valley, 39, 50, 61, 139, 174, 175*, 179
Rhodes, 28*, 29*, 216
Rhodope Mountains, 241
Rhöndorf am Rhein, 172*
Rhone Valley, 131–34, 132*, 298–99
Ribeauvillé, 143
RIBÉRAU-GAYON, JEAN, 64
RICASOLI, BARONE, 151, 153
RICHARDE, EMPRESS, 143
RICHARD THE LION-HEARTED, 71*, 190, 217
RICHARD II, 73
Richebourg, 105
Riderwood, 262
RIEBECK, JAN VAN, 271
Riesling, 173–74
Rimini, 160
Rio Grande do Sul, 268
Rio de Janeiro, 269
Rioja, 197–98
Riomaggiore, 156
Riquewihr, 141*, 142*, 143*
Rishon-le-Zion, 219
Rochecorbon, 126
Rochecorbon, Square Tower of, 124*
Roche-au-Moines, 129
RONSARD, 134
Rohan, Book of Hours, 17*
ROLLIN, NICOLAS, 102
Roma, 275
ROMAINS, JULES, 127
Roman Castles, 162
Romanée, 105
Romanée-Conti, 105
Romanée Saint-Vivant, 105
Rome, 162
Rosés d'Anjou, 128–29
Rosé wine production methods, 56–57
Roseworthy Agricultural College (Australia), 276
ROTHSCHILD, BARON EDMOND DE, 219
ROTHSCHILD, BARON JAMES DE, 82
ROTHSCHILD, BARON NATHANIEL DE, 85
Rouffach, 142
Roussette de Seyssel, 131
Roussillon, 136–38
ROWLANDSON, 76*
Rubicon, 160
Rüdesheim, 174, 176*, 177, 179*
Rumania, 241
RUFUS, QUENTUS CURTIUS, 228
Rully, 99
Rupertsberg, 179
Russia, *see* Soviet Union
Russo, Mount, 256
Rust, 190
Rutherford, 256
Ruwer, 180–81
Rython, 29*

S

Saar, 180–81
Sacavins d'Anjou, 280
Sack, 191
Sacramento, 259
Saint Aignan, 126
Saint-Amour, 94
Saint-Christophe-des-Bardes, 87
Saint-Emilion, 67, 73, 80, 86*, 87–88
Saint-Etienne-de-Lisse, 87
Saint Eucharias Abbey, 168
Saint-Gengous de Soissé, 98
Saint Helena, 256–57
Saint Hilaire-Saint-Florent, 128

Saint-Hippolyte, 87
Saint John of Jerusalem, 217
Saint John's Mountain, 176
Saint Joseph, 211–12
SAINT KILIAN, 183
Saint-Lager, 94
Saint-Laurent-des-Combes, 87
SAINT MARTIN, 126
Saint Maximin Abbey, 168
Saint Nicolas-de-Bourgueil, 127
SAINT NIVARD, 111–12
SAINT PATRICK, 132
SAINT PAUL, 36
Saint-Péray, 133
Saint-Pey d'Armans, 87
Saint-Satur, 124–25
Saint Simon's Day, 234
Saint-Sulpice-de-Faleyrens, 87
SAINT THERESA, 38
Saint-Thibaut, 123
Saint-Tropez, 135
Saint-Vallérin, 99
SAINT VINCENT, 86, 136
SALIN, GUIGONE DE, 102
Salina, 166
Salonae, 244
Samaria, 219
Samos, 30, 217
Samtrust, 231
Samuele Sebastiani Vineyards, 256
San Benito County, 252
San Bernardino County, 259
Sancerre, 123–26
San Diego, 250, 259
San Falviano Church, 162
San Francisco, 256
San Juan, 266
San Lúcar, 192
Sanlúcar de Barrameda, 195
San Marino, 160
Santa Clara County, 252, 258
Santa Cruz County, 252
Santenay, 100–01
Santiago, 267
Santorin, 216
Santorini, 216–17
Saône river, 99
São Paulo, 269
Saratoga (California), 257*
Sardina, 166
SARGON, KING, 228
Sartène, 147
Saumur, 128
Sauternes, 81, 88–92
Sauvignon, 124
Sava river, 243
Savennières, 129
Savigny, 102–03
Savoie, 131
SBARBORO, ANDREA, 255
Scala Dei, 196
Scharshofberg, 181
Schloss Böckelheim, 182
Schloss Eltz, 177
Schloss Johannisberg, 176–77
Schloss Reinhartshausen, 177
Schloss Vollrads, 174–76
SCHUBERT, ANDREAS VON, 181
SCOTT, SIR SAMUEL, 82
Sea of Azov, 231*
Sebou river, 213
SÉGUR, MONSIEUR DE, 82
Sekt, 172–73
Sellery, 112
Semele, 215
Serbia, 242, 244
Sercial, 205
SEREKHOTPE, 18*
SERRA, JUNIPERO, 249, 259, 271

SERRISTORI, CONTE, 151
SESTIUS, MARCUS, 34
Sète, 138
Setubal, 204–05
SEVERUS SEPTIMUS, 31
SEVIGNÉ, MADAME DE, 122
Sèvres-et-Maine, 129
SHERIDAN, RICHARD, 76–77
Sherry, 61, 191–96
Shiraz, 19
Siberia, 227, 229
Sicily, 26, 165, 217
Siena, 151, 153
Sierra del Macho, 196
Sierre, 187
Sigolsheim, 143
SIMON, ANDRÉ L., 74
Sintra, 204
Sion, 187
Slovakia, 241–46
Smuggling, 74–77
Soave, 159
SOCRATES, 23–24
Sofia, 240–41
Solano County, 256
Solera, 192–94
Solin, 245*
Solutré, 98
Soma, 19
Sonoma, 256
Sonoma County, 250*, 252
Sonoma Valley, 255
SOPHOCLES, 28
Sopron, 239
Sous-le-Dos-d'Ane, 101
South Africa, 271–73, 273*
South Australia, 275, 276
Southwark, 74
Souverain Cellars, 256
Soviet Union, 61, 199, 227–33, 229*, 230*, 231*
Spain, 73, 191–98, 195*, 304
Spätlese, 170, 172
Split, 244–46, 245*
STALIN, JOSEPH, 230
Stanislaus County, 259
Steinberg, 177
STEL, SIMON VAN DER, 271
Stellenbosch, 272
STENDHAL, 102
STEWART, LEYLAND, 256
STRABO, 19
Strasbourg, 140
Stromboli, 166
Sumer, 18
Susak, 245
SWIFT, JONATHAN, 199
Swiss Valais, 50
Switzerland, 184–88, 184–85*, 186*, 187*, 188*, 302–04
Sydney, 275
Symposia, 23–24
Syndicat des Courtiers, 81
Syria, 222
Szekszárd, 239

T

Tabarka, 32*
Tabriz, 224
Taganana, 198
Tain, 132*
Tain l'Hermitage, 59*, 132*
Talca Province, 267–68
Tallano, 147
TALLEYRAND, 108
Talmud, 219–20
Tarapacá, 268
Tarbes, 145

Tarragona, 195*, 196
Tastevin, see Wine-tasting cups
Tastevins de Bourgogne, 280
Taunus Mountain, 167–68, 174
Tavel, 57, 134
Taylor Wine Co., 261*, 262
Taza, 213
TCHELISTCHEFF, ANDRÉ, 256
Teheran, 224
Tell, 209
Templars, 217
Tenerife Island, 198
Terroir, 104–05
Thann, 140–41
Thebes, 18*, 20*, 22, 26*
THÉE, ETIENNE, 252
Thibar, 211
THIBAUT THE CHEATER, 128
Thermopylae, 30
THESEUS, 24
Thessaly, 30
Thrace, 30, 223–24
TIBERIUS, 162
TIBULLUS, 33
Ticino, 186
TITO, JOSIP BROZ, 242
Toasting, 74, 174
Tokaj-Hegyaljai, 234–38
Tokay, 61, 225*, 226*, 234, 235*, 236*, 237*, 238*
Tools, 47*, 60–61, 62*, 63*, 92*, 117*, 141*, 253, 266*
Touraine, 57, 126–27
Tour de Marseus, 185
Tours, 122, 126
Tower of the Witches, 142
Traben-Trarbach, 171–72
Tradeau, 135
Transdanubia, 239
Traz os Montes, 204
Treaty of Methuen, 74, 199
Trier, 168, 180, 182
Trittenheim, 168
Trockenbeerenauslese, 171, 176
Tropea, 164
Troy, 28
TUBAL, 205
Tulare County, 259
Tunis, 211
Tunisia, 211–12
Turckheim, 143
Turin, 154, 156
Turning, 60–64
Tuscany, 151–54
Turkey, 18, 61, 223–24
Turkmenistan, 18
TYRKER, 249

U

Ulling, 84
Undurraga, 268
Union des coopératives vinicoles de Tunisia (U.C.T.V.), 212
United States, 249–64, 305–06
Urartu, 228
URDINOLA, FRANCISCO DE, 271
Urjaani, 230*
Uruguay, 270
Urzig, 182*
U.S.S.R. see Soviet Union

V

Valais, 184, 187–88
Valdepeñas, 196
Valencia, 196
VALLEJO, MARICANO, 250
Valley of the Moon, 255

Valparaiso Province, 267
Valtellina Valley, 158–59
Van, 228
Var, 134, 135*, 138*
Vaslin press, 61
Vaudois, 150*
Vedas, 19
Vence, 136
Veneto, 159
Venice, 286*
Venice (region), 159
Verband Deutscher Naturwein-Versteigerer E. V., 169–70
Verdigny, 125
Vergisson, 98
Vermouth, 156
Verona, 159
Versailles, 98
Vertus, 114
Verzé, 98
Verzenay, 114
Verzy, 114
Vesuvius, 33, 162
Veuve-Cliquot, 118
Vevey, 150*, 184–85*, 187*
Vichy, 122
VICTORIA, QUEEN, 168
Vidauban, 135
Vienna, 189
Vienne, 131
Viertelberg, 182
VIGNES, JEAN LOUIS, 249
Vignonet, 87
Vigneronnage, 95
Viguérie royale du Jurançon, 144
Villa Nova De Gaya, 200*, 201
Villafranca del Panadés, 197
Villers-Marmery, 114
Villié-Morgon, 94
VINCI, LEONARDO DA, 158
Vine leaf, 14*
Vineyards, 32*, 45*, 132*, 148–49*, 186*, 187*, 190*, 195*, 211*, 219*, 240*, 245*, 276*
Viniculture Institute, 142
Vinland, 249
Vinprmo, 240
Vinters' Company, 70
Vinters' Hall, 70
Vintimiglia, 156
Vipava Valley, 245
Viré, 98
VIRGIL, 29, 33
Virginia, 259–60
Vis, 244–46, 245*

Viseu, 204
VITELLIUS, 31
Voiteur, 146
Voivodina, 242
Volnay, 102
Volstead Act, 251
Vosges Mountains, 139
Vöslau, 190
Vosne-Romanée, 105
Vouvray, 123*, 126–27

W

Wachau, 190
Wachenheim, 179
WAGNER, PHILIP, 262
WALDEMAR, KING, 70
Washington, 259
Weapons, 147*
Wehlen, 180
Wellington (South Africa), 272
Wenkel, 177
WENTE, CARL, 258
Wente Brothers, 258
Western Europe, 150–206
White wine production methods, 54–56
Widmer's Wine Cellars, 260*, 262
Wiesbaden, 174, 177
WILLIAM OF ORANGE, PRINCE, 176
Willmes Press, 253
WILSON, WOODROW, 251
Wine, composition of, 62–63
Wine, history of, 15–39, 67–77
Wine appreciation, 279–93
Wine blending, 79, 116
Wine bottles, 9–10*, 83*, 243*
Wine brokers, 79–80
Wine buying, 78–79*
Wine cellars, 60*, 89*, 106*, 176*, 179*, 197*, 257*
Wine classification, 80–82, 93–94, 117, 169–70, 174–76, 185–86
Wine cups, 28*, 30*, 39*, 158–59*, *see also* Wine glasses; Wine-tasting cups
Wine Export Act of 1924 (Australia), 275
Wine festivals, 22–23, 23*, 26*, 31, 102–03, 143, 150, 179–80, 184–85*, 201, 216, 265*, 269–70
Wine flasks, 153*
Wine glasses, 104*, 282–83*, 285*, 286*, 288*, *see also* Wine cups, Wine-tasting cups

Wine guides, 76
Wine jars, 223*
Wine labels, 118–19*, 154–55*, 165*, 166*, 172*, 173*, 199*, 205*, 210*, 232–33*, 248*
Wine lists, 207*
Wine making, 20*, 28, 33, 37*, 38, 51–65, 60*, 65*, 82–84, 88–92, 89*, 90*, 92*, 101–102, 114–21, 130*, 133–34, 144–46, 158, 171–72, 187, 191–94, 201, 205, 216, 231*, 232–37, 253*, 254, 258*, 261*, 263*, 266–67, 270, 274*, 275*, *see also* Grape growing; Harvesting
Wine-making tools, *see* Tools
Wine presses, 60–61, 62*, 63*, 253, 266*
Wine production, 308–09*
Wine-cooking recipes, 293–306, 295*, 303*
Wine sales, 74, 102
Wine selection, 80, 289–90
Wineskin, 72*
Wine tasting, 77–79, 78–79*, 89*, 94*, 97*, 126–27, 178*, 179*, 206*, 230*, 279–88
Wine-tasting cups, 66*, 102*, 109*, 277*, *see also* Wine cups; Wine glasses
WOODHOUSE, JOHN, 165
Worms, 178, 181*
WUILLOUD, 188
Würzburg, 183
Würzburger Stein, 183

Y

Yarilo, 229
Yeast, 57–58
YI TIEH, 19
YU, EMPEROR, 19
Yuan Dynasty, 19
Yugoslavia, 242–46, 242*, 243*, 245*, 246*
Yvorne, 185

Z

Zell, 167*
Zellenberg, 143
ZELLERBACH, JAMES D., 255
Zeltingen, 180
ZEUS, 22, 215
Zichron Jacob, 219
Zupa, 244